John Balbirnie, George F. Adams

The Physiological Basis and Curative Effects of the Turkish Bath

John Balbirnie, George F. Adams

The Physiological Basis and Curative Effects of the Turkish Bath

ISBN/EAN: 9783337290795

Printed in Europe, USA, Canada, Australia, Japan

Cover: Foto ©berggeist007 / pixelio.de

THE PHYSIOLOGICAL BASIS

AND

CURATIVE EFFECTS

OF THE

TURKISH BATH;

BY

JOHN BALBIRNIE, A. M., M. D., (LONDON).

REPUBLISHED BY

GEORGE F. ADAMS, M. D., OF ST. LOUIS,

Late Professor of Minor Surgery and Medical Jurisprudence in the "Female Medical College," Boston, Mass.

CONTENTS.

I. *Preface.*—By Geo. F. Adams, M. D.
II. *Treatise.*—By John Balbirnie, A. M., M. D.
III. Opposition to New Truths, a characteristic of the Medical Profession.
IV. Col. T. T. Gantt's Lecture to the Citizens of St. Louis.
V. The Turkish Bath, from the St. Louis *Medical Archives.*
VI. Physiological Superiority of the Turkish Bath over the Russian Bath.
VII. Interesting Account of the St. Louis Turkish Bath.—*Missouri Rep.*
VIII. What the Turkish Bath will do for a Person.—*Missouri Democrat.*
IX. Testimonials from Patrons of the Bath.
X. Turkish Baths Versus Arkansas Hot Springs.

PRICE, 25 CENTS.

First Edition Five Thousand.

TO THE PUBLIC.

My object in republishing this little work, by Doctor Balbirnie, with additional testimony appertaining to the benefits of the Turkish Bath, is to lay before the public some authentic and reliable facts concerning this important and interesting subject. The Bath in this country has, comparatively, but few advocates among medical men, and consequently no history of its own. In Europe and on the continent it is far different. There many of the acknowledged best writers and scientific men, of all schools of medicine, have written much upon the subject and sustained the principles, theoretically and practically, of the hot air bath, and have introduced it as a remedial agent into the oldest and most popular hospitals and asylums in England, Ireland, Scotland and Germany. The reader will find the subject so ably handled by Doctor Balbirnie that no words of mine are required to urge a thorough perusal of the same. I have also added a few testimonials, with other facts, in regard to the bath under my superintendence here in St Louis, which I trust will not be altogether uninteresting to the reader.

On the 9th of October, 1869, this bath was opened to the public. The encouragement we received from many of our best physicians (of both schools) was very flattering, and to the kindness and liberality of these gentlemen I owe much of my success, for which, I trust, I am duly thankful. Since we opened the baths (eighteen months ago), our register shows that more than twelve thousand persons have taken them in this institution.

To say that I have received the encouragement from the Faculty I had every reason to suppose I should, would be untrue. While many of our very best physicians have used the bath, and sent many of their patients here from time to time, others have denounced it as a humbug, a dangerous operation, a weakening process, etc., and attempted to keep people away by ridiculing the thing as a catch-penny affair, and some of the *would-be shining lights*, have gone so far as to advise their patients to make their wills before taking a bath. To those gentlemen a cordial invitation is again tendered to avail themselves of the first principles involved in the administration of the Turkish Bath.

I am daily interviewed by my patrons as to what the Faculty think or say as to the remedial effects of the bath? That question is well answered by Dr. Erasmus Wilson, of London. He says: "When you present any simple means of general application which will prevent or cure a large class of diseases, the doctors, very naturally, divide on the subject. The real scientific men, those who regard the interest of humanity above the gains of their calling, behold in it a great instrument of good, and thank God for it. The practitioners of the opposite class behold in it a great enemy, and treat the matter accordingly."

Dr. Hanson, of Milwaukee, an able writer and a thorough gentleman, says: "The recognition of the Turkish Bath, as the chief handmaid of the true science of Medicine by the practitioners of all schools, is only a question of time. Science and experience commend it, and the people will, as thousands have done, patronize it, either by the advice of or in antagonism to the views of their physician, and policy will convince all those whom better reasons fail to reach."*

The public will think and act for themselves by and by, as the salutary effect of the bath is not a matter to be judged by any doctor's opinion. *It is a matter of simple evidence*, concerning which the intelligence that rules a "common jury" is just as competent to judge as all the colleges of physicians in the country combined, and, indeed, with a vast deal stronger probability of an unbigoted and impartial verdict being returned.

There is much ignorance prevailing on this subject (much of it wilful, I fear), and so much said about the bath being a new thing, another nine days' wonder, gotten up to humbug the people, &c., &c., that I propose to go back a few years and see what history has to say in regard to the hot air bath. In the reign of Antoninus Caracalla, A. D. 302, Gibbon says: "The baths of Caracalla were opened at stated hours for the indiscriminate service of the senators and the people; that they contained about 1,600 seats of marble, and that the Thermæ could accommodate more than 3,000 persons at one time. They occupied the 'Aventine Mount,' and excelled in beauty, grandeur and extent those of any former Emperor. The whole enclosed space was more than one mile in circumference; the total length of the Thermæ, or hot air chamber, was 1,840 feet, and its breadth 1,476 feet. At each end were two temples; one dedicated to Appollo the other to Æsculapius, as the tutelary deities of a place sacred to the improvement of the mind and health.

"The baths of Diocletian excelled even those of Caracalla in extent and splendor, and were the largest in Rome, or, indeed, in the world, for they were capable of accommodating 18,000 bathers at once. According to Eusebius, they were completed A. D. 302, and were built, principally, by the enforced labor of Christians, during the tenth and last persecution."

All the baths of the Emperors had the air heated by flues underneath the floor—the hypocaustium---after the Greek model. So highly valued was the bath by the military authorities, as a sanitary institution, that wherever a permanent Roman camp was formed there also baths were constructed, to protect the health of the soldiers. The extensive remains of several such have been discovered in England, at London, Chester, Weaxeter, and elsewhere.

* The above was written by Dr. Hanson four years since. To-day his prophecy is fulfilled (at least in the little city of Milwaukee), his daily average of baths being more than one hundred.

Candidly considering the history of the bath, the conclusion irresistibly forced on the mind is, that its utility alone was the cause of its universal adoption, for it would be ridiculous to suppose that it could possibly have survived the vicissitudes of ages had it not possessed *a healthful and curative potency*, which commended it to *the practical wisdom of mankind.*

As an institution, valued and honored, the bath flourished in the most renowned nations of antiquity—the Assyrians, Persians, Egyptians, Greeks and Romans; and when it was buried amid the splendid ruins of the Greek Gymnasiums, and shared the destructive fate that overwhelmed the resplendent architectural glories of Imperial Rome, it nevertheless survived in far distant parts of Asia.

The universality of the bath, therefore—its existence during thousands of years in every quarter of the globe—furnishes prima facie proof of its value far superior to any that mere abstract reasoning could supply. It might have been imagined that such an institution would naturally commend itself to a Medical Profession, conscious of its high, moral obligations, if not suddenly at first, at least *after* some of the most eminent medical men of the day had tested its merits and certified to its great therapeutic power. But the bath has experienced no appreciative welcome from the Profession, as a body, more particularly from the teachers and leaders of medical opinion. "We ask ourselves," says Erasmus Wilson, one of the first authorities of the day, "not what disease will be benefited by the bath, but what disease can resist its power?" Yet, notwithstanding a "cloud of witnesses" have given similar testimony, those who control medical teaching, opinion and practice still remain obdurate; and, taking the most charitable view of the matter, I can see no reason, neither can I find any cause, for the apathy that prevails in the Profession—an apathy which amounts to inhumanity. An agent, at once safe and powerful, agreeable and economical, is offered, fully tested by experience, and certified as incomparable in relieving various phases of the most painful diseases, and yet how many physicians can give an intelligent answer to his patient when questioned as to the principles of the hot air bath, their effects, &c., &c., upon diseased action. I assure the profession their time can not be spent more profitably than by giving some portion of it to this subject.

And now, in offering this able and useful work, on the "Heat Cure," by Doctor Balbirnie, I can only say that I hope much good will come of it, and that many who now condemn and ridicule the bath. both by words and works, will see their way clear to change their views and overcome their former prejudices sufficiently, at least, to give the bath a personal test, and also to give the following pages a thorough perusal. My object in republishing this work will then have been accomplished.

GEO. F. ADAMS, M. D.,

1603 *Washington Avenue, St. Louis, Mo.*

PREFACE TO THE FIRST EDITION.

The question which we here feebly essay to expound is something more, and higher far, than the introduction amongst us of an Oriental luxury, a pure custom, or a new mode of cleanliness, all important as it is admitted to be, even in these subordinate points of view. THE TURKISH BATH IS A MIGHTY AGENCY FOR THE PREVENTION AND CURE OF DISEASE. It is a long sought *desideratum* of practical medicine, and will be hailed by all physiologists and physicians (who have more at heart the interests of humanity than the gains of a calling) as one of the most potent *modifiers of the living organism*, whether in health or disease. In this aspect of the subject, the more this new ally of the Healing Art is tested the more it will be trusted—the more will it vindicate its pretensions to be placed in the arsenal of physic, as an *orthodox weapon* of medical warfare. As such we believe the day will come when its machinery will be established as an integral and essential part of the equipment of every hospital, dispensary, asylum, workhouse, barracks and camp—yea, even of every public school, of every civilized State. Increasing experience is bringing forth new facts every day to prove its curative powers.

Will our palaces and metropolitan club-houses be long without the Bath? We trow not. How long will it be ere it becomes the health-preserving implement of every complete private mansion? No other agency will so neutralize the drawbacks and discomforts and dangers of our cold, damp, variable climate during at least seven months in the year. Whatever may be alleged of the *curative* powers of the Bath, it can not fail, bye and bye, firmly to establish itself in the public confidence as the grand PROPHYLACTIC of disease —the PREVENTIVE agent *par excellence.*

There can be no question but that the Turkish Bath, extensively put within the reach of the poor, will do much to supplant the baneful fascination, and to substitute the injurious stimulation of alcoholic liquors! It will become, perhaps, the most powerful antagonistic or counteractive agent the Temperance Cause has yet wielded. That sacred cause must seek, as its three grand allies in exalting debased humanity, Cleanliness, Health and Religion — and *the accredited ministers of those agencies.* The most speedy and summary way to put down the nuisance and demoralization of the GIN PALACE will be to *pit it against and to pitch against it* a Turkish Bath of at least equal decorative attractions—and offering to the poor, for the price of the poisonous dram, two hours' oblivion of their care and misery—with improved health, quiet nerves, natural appetites, and, perhaps, washed raiment at the conclusion of the process. A soup kitchen or a working man's refreshment room will be a necessary appendage to all such establishments. It will require no gift of prophecy to predict which place of resort shall receive most patronage, and how far the improved feelings, and thoughts and habits so induced will pave the way for the labors of the city missionary. Will not some wealthy philanthropist, or society of philanthropists, try the experiment? Will not the teetotallers take up this question?

It is, perhaps, not out of place here to allude to, to demolish a prevalent misapprehension on the subject of the Turkish Bath: it is supposed to be only suitable for strong constitutions! This is a complete mistake. The weakly, to the contrary, as they have more need for it, are, perhaps, more benefited by it. Its influence as an instrument of *training*—as a means of physical development—is the least questioned and questionable. Powerfully aiding NUTRITION, it manifestly promotes growth and strength. For all, therefore, in whom nutrition is depraved or defective—for the scrofulous, the consumptive, the ill nourished, the enfeebled, the emaciated, &c., the Turkish Bath is pre-eminently adapted. Nor is any extreme of age beyond its scope. Indeed, the national use of the bath, for ages, by the Persians, Greeks, Romans, and (since the conquest of Constantinople) by the Ottoman nations, demonstrates, at least, the utter groundlessness of its alleged dangers.

But every excellent thing, even the best, may be abused. The Turkish Bath is too powerful an agent for good not to be an equal instrument of evil *when misapplied*. *Its* dose requires to be regulated like that of any other remedy—and this certainly is the province of the physician. To be wielded, therefore, with safety, precision and success in the treatment of disease, and for the invigoration of the delicate—to be delivered from the evils of its maladministration, and to prevent such accidents as have already occurred in this country—to save, in short, a good cause from a bad name, the Turkish Bath must be under scientific prescription and skilled superintendence.

In conclusion, it may be affirmed that the Turkish Bath amounts almost to a DISCOVERY! It is at least a new found boon to the States of the Western World. We claim for it to become a permanent institution among them, as a remedy for many of the evils of modern civilization—a remedy near at hand, safe, effective and agreeable. The questions it stirs are those which, next to morality and religion, intimately affect a nation's best interests. The habits it promotes are those which most directly conduce to the health, the happiness, the longevity, the physicial culture, the material prosperity, and the moral elevation of the people.

JOHN BALBIRNIE.

CLAREMONT HOUSE, GREAT MALVERN, May 8, 1863.

PREFACE TO THE SECOND EDITION.

In issuing a new edition of this essay, the Author regrets that the most of it has been worked off and stereotyped while absent on a tour. This has prevented both press corrections and the addition of new matter necessary to perfect the physiological *rationale* of the Turkish Bath. It strikes the writer that his own, and all other explanations of its action and virtues have been too *mechanical*—have been founded too much on what might be termed the SCAVENGER WORK of the Bath—its *safety valve opening and drain flushing* operations. Undoubtedly this is a true and and all important point of view; and *alone* would place the Turkish Bath on a high pinnacle of pre-eminence, not only as a means of cleanliness and luxury, but as an instrument of Therapeutics. To macerate the corporeal tissues and thereby to soften and open up their porous structure, obstructed by disease, by sedentary occupations, or by luxurious modes of living; to clear off the epidermic varnish that mars the *breathing functions* of the skin, to exalt its exhalant and absorbent powers, and thereby to enhance its uses as a prime agent of the aeration and CIRCULATION, as well as of the purification of the blood; to set free the blocked up excretions of the body by clearing their eliminatory outlet, and thus to facilitate what is called "the metamorphosis of structure;" by powerful, yet unweakening perspiratory drains, to equalize the distribution of the blood on the surface and in the interior, and thus to undo congestions of vital viscera. Simultaneously with all this, to *poultice* (as it were) the extremities of the nerves, to soothe the sentient external surface, thereby most effectively quelling internal irritation and quieting brain excitement; and finally to close the patulous pores and brace the relaxed muscles; and then virtually to *electrify* the whole system by the finishing-off ablutions; certes, these are grand ends to gain—an immense boon to the sick or the sound man; and these, moreover, are the express aims and "indications" of all medical practice, by whatsoever name called. Thank God, the sure accomplishment of these ends is the valid boast—we had almost said the exclusive prerogative—of the Turkish Bath! So far we can point to its *demonstrable* sphere of action. This is, however, the utmost length that writers have hitherto gone in their appreciation of the *modus operandi* of the Bath. But these effects, how valuable soever, are after all in a sort merely *mechanical*, and constitute only one half, perhaps the least potent half, of the physiological benefits of the Bath. There is something more and greater far beyond—something "behind the scenes," though less palpable, yet paramount. We have now to unfold *vital* actions of a higher class than the results specified—actions which it is the aim of all medicines to effect; and which the very best medicines, by a rare chance only, succeed in effecting.

That copious visible distillation of fluids from the skin has its precise counterpart and analogue in the excretory actions taking place within—on the mucous, and even from the *serous* surfaces—from the ducts, and even the parenchyma of glands, perhaps even from every capillary tube and strainer. In this grand interior physiological *molimen* taking place always under the operation of the Bath, is to be sought the explanation at once of its invigorating and of its curative powers.

As elucidating the philosophy of EXCRETION, or the DEPURATING ECONOMY of the body—we have on page 22 referred, in brief, to the beautiful physiological doctrine of CELL-FORMATIONS—minute vesicular bodies, wherein all the chemicovital actions of the organism are effected. By the growth, filling and bursting of these nucleated cells, all absorpion, all secretion, and all nutrition are performed. We were content there with a mere allusion to the subject. But as it constitutes in a sort the very *key of the position*—the stronghold of the fortress of truth—the Bath partisans contend for, the subject must be opened up at greater length, and illustrated and enforced so far as limited space will allow.

Both the *organising* and the *disintegrating* acts of secretion are examples of the beginning and the ending of the CELL-LIFE now in question. The favoring conditions for the development of cell-action, when the dormant or latent germs of it exist, are HEAT and the PURE OXYGEN of the atmosphere. Instance the case of the growth of the chick in *ovo*, or of the seed in the soil, even if that seed has lain 3,000 years in the coffin or stomach of a mummy! Here *cell development* or secretory action in the *fons et origo* of the formative *nisus*, and not only the beginner, but the maintainer and the ender of it—till the "topstone" of the animal or vegetable structure is put on! "How is all this proved?" the unphysiological reader asks. We reply, "The microscope has brought to light these dark arcana of nature." Let this reply so far suffice for the present. No instructed medical man will think that in treading this ground I am going out of my way for material of defense of the Turkish Bath. The phenomena of the gardener's *hot-house* (whether it be in the way of developing almost at will, foliage, flowers or fruit, *or whether in keeping in health and vigor tender exotics that our rude clime would be fatal to without such fostering care*)—I say these familiar phenomena are illustrations of our control over CELL-ACTION. The sights in our prize cattle exhibitions show our control of the SECRETORY ACTIVITY of *animal* organisms, pushed even to a *morbid* excess. The simple agents at work here, in addition to the *nutrient materials* (which must, in all cases, constitute the platform of operations), are temperature and pure air. Of course, VITAL ACTION is, above and beyond all, THE CONTROLLING POWER. But the grand point to insist upon is, that this very supreme vital action is itself under the control of Art! Those SECRET, SECRETORY, FORMATIVE PROCESSES which we can initiate and evoke at will, as in the chick or seed, or which we can control in the animal as to produce all modifications of blood, bone, nerve, vessel, brain, muscle; or by which, in the case of plants, we vary at pleasure, roots, stems, branches, leaves, buds, flowers, fruit---these, I say, ARE PRECISELY THE SAME PHYSIOLOGICAL ACTIONS WE CALL POWERFULLY INTO PLAY IN THE TURKISH BATH. This has not been laid due stress upon---if the point has been mooted at all! and we are not aware that it has! A thousand facts prove that the caloric and oxygen of the air, largely received by every pore of the skin, and every vesicle of the lungs, start into unwonted activity the processes of *cell-development or secretory action.* This is the basis and beginning of all salutary, life-exalting, disease-curing efforts on the part of the organism. Of course, the subsidiary agency of diet and regimen, air, exercise and repose must be invoked and scientifically regulated. But in virtue of this secret physiological machinery of cell-operations---a true secretory *nisus*---it is in our power often, suddenly and at once, to extinguish the disease, and reconstitute, rebuild and re-energize the dilapidated and decaying bodies of our fellow men---if slowly, sometimes, and by a very bit-by-bit process---yet unfalteringly, and without failure as without check, *in the same way, by the self-same mechanism, as the coral insect (out of its secret infinitesimal secretions) piles up its rock-reef or sea-girt isle!* After these palpable and pertinent instances of CELL-ACTION, who shall attempt to call "romancing," or to think incomprehensible, incredible or mysterious the infinitesimalism of nature's operations, or to question the grand results they achieve. "*Si le grand Dieu est grand dans les grandes choses, il est tres-grand dans les petites.*"

Space forbids us here to pursue the subject; we have thrown out sufficient hints for the reflective. I have given the clue to the true *rationale* of the best results of the Bath.

JOHN BALBIRNIE.

THE TURKISH BATH, SOUTHPORT, July 30, 1864.

THE TURKISH BATH.

CHAPTER I.

THE PHYSIOLOGICAL BASIS AND ACTION OF THE TURKISH BATH---THE PHILOSOPHY OF DEPURATION---THE PENALTIES OF NON-DEPURATION.

"THE BLOOD IS THE LIFE," as charged with the great vital STIMULI, *i. e.*, the sustainers of the movements of the animated machine, the sources of its heat, and power, and action; as containing, on the one hand, the elements of nutrition, or the building materials of the fabric—and the fuel of the living furnace; and on the other, the atmospheric oxygen necessary to ventilate the house we live in—to combine with the products of decomposition; thus, in one act, by one process, supporting the combustion of the body, keeping up its heat, and effecting the removal of its waste. This waste is better understood under both its popular and its scientific name---the *excretions*, or the skimmed off impurities of the body. EXCRETION is, therefore, the depurating process of animal bodies, which we must, if possible, enable the reader fully to understand, if he is to comprehend the action and appreciate the virtues of the Turkish Bath.

The mere functioning or play of organized structures, every movement, great or little, of the living apparatus, even every act of volition, every thought and every emotion, disengages heat and dissipates it, and, therefore, by the first laws of chemistry, must wear down and disintegrate the mechanism piecemeal. Hence from its first development to its final dissolution, the body is in every atom (especially of its soft parts) the scene of incessant, even of momentary CHANGE---of Reproduction and Decay---of the displacement of the molecules of the old and effete matter, and their combination in new forms, in order to their exit from the body.

The healthy properties of the living fabric are maintained only so long as a due equilibrium exists between NUTRITION and EXCRETION, or DEPURATION; in other words, between *supply* and *waste*---between *income* and *expenditure* of body elements---between the ASSIMILATION of the new materials and the ELIMINATION or exit of the old, worn out, or superfluous constituents of structure.

In the outgoing rounds of the circulation (*i. e.*, by the arteries), the blood yields up its nutrient principles for the growth or repair of the several tissues; in the incoming or returning circuit (*i. e.*, by the veins), it receives for removal or revivification, the particles that have been exhausted of their vitality, or that have served their purpose in the economy. This corporeal *debris* (*sewage*) imparts to the blood a dark color and poisonous properties. Hence the great importance ever attached to keeping in good working order the EXCRETING APPARATUS of the body. This was the grand virtue of Old Physic (which we willingly concede to it), of giving minute attention to the *excretions*. The aim was right---the means faulty. Irritant medicines *create* the anomalous excretions they were supposed to eliminate.

Secretion and Excretion are often used as synonymous terms. They mean the same thing---literally something *separated* from the blood for specific purposes in the living mechanism. 1st. For the preparation of the nutrient materials, as the saliva, gastric and pancreatic juices, bile, &c. 2d. For the formation of the solids and fluids of the body, as bone, muscle, nerve, tendon, the serous fluids of the joints and of the "shut sacs," the humors of the eye, tears, mucus, &c. 3d. For the straining off and outlet from the system of all substances whose retention would be injurious---all wasted, extraneous, or superfluous matters. These latter constitute the EXCRETIONS PROPER. The excretions are to be viewed as the living *waste pipe apparatus* for equalizing, as nearly as possible, the availing amount of the body's reparative materials to the degree of its wear and tear. The excreted products of the body, therefore, are, or should be, equal in amount to that of the solids and fluids ingested.*

The DEPURATING PROCESS of animals is more essential to life than even nutrition. There is but one apparatus or system of organs, and that comparatively a small one, appointed for the elaboration of the food; but many and large are the instruments appropriated to the extrication---the *excretion*---of corporeal waste. The LUNGS, LIVER and SKIN are set apart for the elimination of the effete or superfluous *carbon*. The KIDNEYS are the grand outlets for the decomposed *nitrogenous* matters, and the earthy and saline materials. Every other function may be suspended for a considerable time without involving life. We can live for weeks without food, or with the liver "locked up;" and several days with the functions of the.

*A practical reflection here. A man, if he suspects his state of health, may thus summarily test it, take disease "by the forelock," and save himself much after-suffering, by simply asking, "Is my legitimate waste in labor or exercise equivalent to the quantity of good things I daily consume; and are there no *capillary obstructors* among those good things?" If the answer of conscience, or intelligence, or experience, as to these points, is unsatisfactory, then is his "nick of time" to diminish or cut off the supplies, and to hie him to the taking-down, swilling-out and rinsing-off process of the Turkish Bath.

"PRINCIPIIS OBSTA: sero medicina paratur,
Cum mala per longas convaluere moras."

How many valuable lives would thus be prolonged? How much invaliding prevented? How much medical practice superseded? Is the physician philanthropist enough to rejoice hereat!

kidneys annulled; but we can live only two or three hours with the skin coated over, and only a very few minutes with respiration suspended! HENCE IT IS CLEAR THAT THE INTEGRITY OF THE ELIMINATORY, OR DEPURATING, FUNCTIONS IS THE FIRST WANT OF ANIMAL LIFE—THE INDISPENSABLE CONDITION OF SOUND HEALTH. From the same facts, as well as from the immense extent and influence of the LUNGS and SKIN, it is very manifest that the *grand business of* DEPURATION *falls chiefly on these organs.*

The EXCERNENT, or DEPURATING, ACTIONS and APPARATUS of the living organism it behoves the lay-reader well to comprehend, if possible. In their philosophy lies the basis of all explanations of either the Theory or the Practice of the Healing Art. As elucidating this question, we must here devote a sentence or two to the subject of CELLS—the secret, retired, infinitesimal organisms, which are the true builders of all animated structure. Every vitalizing act commences in CELLS. Nutrition and secretion, growth and renovation, are but a series of *cell* operations! Fat is thus *excerned*, separated from the blood, in its little bags (ADIPOSE TISSUE). Glandular secretions are but the bursting and yielding up of the contents of the CELLS covering membranous surfaces, or lining the follicles and tubes of glands. The mucus, which coats the surface of the mucous membranes, is elaborated by *epithelium cells.* The *epidermis* (or scarf-skin) is but another form of these cells, their contents dried up and exfoliating. The cells are continually developed, cast off, and renewed from the germs supplied by the subjacent membrane. The CELLS of the intestinal VILLI (pile of tufts) select and separate from the contents of the alimentary canal the nutritious from the refuse matters. In like manner, the CELLS of the secreting tubes, follicles, or passages of a gland (as the liver, the kidneys, &c.), separate from the blood the effete matters it is its function to elaborate and discharge (as bile, water, &c). ORGANIZATION is simply the appropriation thus of the nutrient compounds floating in the blood, and their combination in the proportions necessary to produce all the diversified "tissues" or structures of the body—here bone, there brain; here muscle, there mucus; here nerve, there vessels, &c. This organizing process is sometimes called ASSIMILATION—a vivifying or life-giving process; assimilation is literally making food *like to*, or part and parcel of, the tissues.

From all the above, it will be clear that the presence of any unassimilable matters in the blood—substances foreign to nutrition—as drugs and other poisons, miasms, the *ova of entoza*, *&c.* or sheerly *the unremoved waste of the body;* in other words, RETAINED EXCRETIONS—will risk the elements (*e. g.*, *hydatid sacs*, or *cistircirci cellulosi*). In this way worms are found in the brain, "flukes" in the liver, &c.; cancerous tumors are developed, and deposits of tubercle formed, &c. In the same way, we have to explain the local irritations, the pains, the functional disturbance of organs, the deteriorated nutrition, the decline of strength, and the constitutional suffering attending the course of certain diseases. In short, the alterations in the body effected by the loss of balance between the functions of nutrition and depuration—the retention, or retarded elimination, of the products of decomposition—or foreign substances accidentally or voluntarily introduced—lie at the foundation of most diseases, and constitute their most palpable material conditions. The mere reactions taking place between the solids and fluids of the body, *in channels where the circulation is barred* (*e.g.*, in congested viscera), suggests, even to the lay mind, sufficient cause of deranged health, malaise, and misery. Imagine only half-an-inch of the finest hair—an eyelash—dropped in among the machinery of a Geneva watch. The living organism is, beyond all comparison, more nice and complex, and, at least, not a whit less sensitive to disturbing causes!

Healthy blood-making depends infinitely more on perfect depuration—that is, on the active condition of the excretory functions—than on the abstractly nutritive qualities of the food. Whenever the body's *debris*, or the matters of its decomposition, are not duly excreted, a virtual and valid *materies morbi* remains to vitiate the process of recomposition. The functions of supply being impaired—the fountains of corporeal renewal being tainted—the *educts* and *products* of the assimilative process must be faulty. Bad materials can only furnish bad building. Hence the commencing loss of high *condition* whenever man comes materially to infringe the Hygienic laws; when, for example, superfluous food and pernicious drinks combine, with the want of due activity of the *lungs* and *skin* (*i.e.*, with corporeal inaction), to derange the balance between waste and supply. Even the diet may be proper as to quantity and quality, and the alimentary canal may be kept "clean;" but all will not avail to produce healthy blood or firm textures, so *long as the pulmonary and cutaneous safety-valves are obstructed or marred in their play.* Let me, however, here remark, so intimate are the connections and sympathies between the Skin, Lungs, Liver, and Bowels, that, under the circumstances described, it is impossible to keep the alimentary canal "clean," even in the sense which leaves out of view the operations of digestion. Those who feed the best, in the popular acceptation of the term, are not the best nourished. An inferior aliment will be turned to good account—any ungenial substance it contains will be neutralized, strained, or *burnt off*, provided the air breathed, and the exercise taken, by the individual, be such as to keep up a highly active state of the grand eliminatory outlets of the body; in other words, *provided the* LUNGS and SKIN *have the fullest scope for the performance of their appropriate functions.*

Here it falls into place to illustrate the effects of inactive *depuratory organs*, from sedentary habits, indolent repose, and luxurious indulgences of all sorts. The structure and functions of man show that he was not by any means intended to be a *sedentary animal.* Those who live the longest and enjoy the best health are invariably persons of active habits. *From the moment man becomes a civilized being, the Depurating process of his blood becomes less perfect; in other words, the grand excretory functions of the Skin, Lungs, and Liver are less completely exercised. From that moment, diseases of various type and class, and one large class in particular—tubercular diseases (scrofula and consumption)—begin to show their ravages on his frame.* And the reason of this is very obvious. Man's habits and modes of life become then less conformable to the instinctive requirements of his constitution; his exercise is less frequent or less natural—either unremitting or not at all; his lungs are compelled to long periods of comparative inactivity; and his skin is equally diminished in function by loads of superfluous clothing, as well as made susceptible to every atmospheric variation by all sorts of "coddling" in warm rooms. By these and sundry other anti-hygienic influences, the *blood of the civilized man is infinitely less oxygenated than it should be.* He voluntarily debars himself of the means of carrying off the effete matters of his body. When the Lungs are imperfectfy exercised it is impossible for the skin to be healthily active in its duties; for the two go together, co-functionate (if we may coin a word). Baths (of the old sort) and cleanliness were the best compensations the case admitted. But nothing—*save some such substitute as that now presented to the public in the shape of the Turkish Bath*—perfectly compensates the want of active exertion in a pure air; for nothing else can perfectly open, and keep open, the body's safety-valves, or secure the perfect elimination of the corporeal waste.

But the worst of the case of the locked-up excretions of the *skin* in particular is this—viz., *that the duty so shirked is thrown necessarily on the Lungs, Liver, Kidneys, or Bowels.* Hence, *nolens volens*, able or unable, the latter organs *are compelled to do double work!*—viz., to perform their own specific work, and to take up the superseded and suspended functions of the skin. For a time, the constitution gives no indication of the injury of this supplementary labor or vicarious discharge of duty. But eventually, at "the turn of life"—at the critical age—in short, in the period of decline, the overtasked, overstrained organs, *knock under.* Nature tires, or gets deranged, in the unequal conflict. From this starting point, a chain of morbid causation gradually stretches its links round the organism, first impeding, then disabling function after function. The liver or kidneys utter their complaints with a voice that can neither be misinterpreted nor resisted. Congestion of the abdominal viscera is imminent, and blue pill is at a premium; or diuretics or cathartics are in demand. The heart, lungs, or brain, show open and manifest signs of congestion, at least of embarrassment and tardiness in their operations. The individual ages rapidly. His face is tallowy or jaundiced. He is the victim of sciatica, or TIC, or gout, or rheumatism. In short, from a primarily inactive skin, *aided by an overactive or over-stimulated stomach, and perhaps an over-worked or over-worried brain*, the sufferer becomes prematurely old and regularly broken down—a victim of disease too generally incurable, involving the principal organs—all the product of impaired general, and arrested local, circulation—congestion of vital structures—and, with this state of matters, retained excretions, and poisoned life-springs; all—all from the simple starting-point of UNDEPURATED BLOOD! A host of evils, therefore, in their beginning, perfectly subject to man's control, and within easy reach of remedy.

The most desolating, as the most universal, scourge of modern society, viz., TUBERCULAR DISEASE—has its origin *in impaired functions of the skin and lungs!* This is usually supposed to be purely and simply a disease of disordered nutrition. *And so it is in all its essential elements.* But the *fons mali* lies farther and deeper. Neither digestive derangements nor scanty nutrition ever, *per se*, generates this "foul fiend." Be it thoroughly well considered and remembered, it is only *when impaired nutrition, or bad blood-making*—whether from bad materials or bad stomach—*coincides with forced inaction of the pulmonary and cutaneous functions*—that is, with defective elimination of carbonic acid and lactic acid—that the dire blood-taint in question, and its *characteristic products*, are manifested. Multitudes of scrofulous and consumptive patients do not belong to the ill-fed classes, neither are they among the notoriously dyspeptic; or they only become dyspeptic in the advanced stages of the malady. On the other hand, it is a matter of familiar observation that your thorough-going dyspeptic—and his name is Legion—never becomes either scrofulous or phthisical. As a general rule, he is a being who lives very much for himself, and therefore with extreme care—one who encompasses himself with the comforts of life—who eschews excesses, and who has a care to breathe pure air—who takes much exercise, who bestows much pains on the condition of his skin, giving it every advantage of clothing, cleanliness, currying, suitable temperature in doors, &c. Besides, your gastric sufferer is usually a keen man of business, or an ardent devotee of literature and science, and is not devoid of much agreeable mental stimulation. *All these are conditions opposed to the inroads of tubercular disease!* But let the circumstances of the case be reversed—let the individual be ill-fed, ill-warmed, ill-housed, ill-clad, ill-ventilated—let him become the inmate perhaps of a cellar residence, or a prison cell, with *moral* as well as *physical* depression, low spirits, &c., to struggle with—and it will then be a miracle if he do not, sooner or later, exhibit some form of this exterminating disease. But the morbid change in question (*tuberculosis*) takes place, less *because* of the implication of the digestive organs, than because the lungs and skin have been condemned to comparative, if not absolute, inactivity. The very sighing of the disconsolate is an instinct to rouse the action of the lungs. In like manner the well-to-do classes, who have no material or ostensible miseries to borrow the diease from, equally succumb, when blighted affections, grief, bereavements, disappointments, &c., deaden both heart and head, paralyzing, in a sort, the skin, and lungs, and liver, if not limbs also. *In short, any one may become the mark and* *victim of tubercular disease, when, together with causes impairing the general health*, the active play of the skin and lungs is impeded, *from any circumstances whatever.* The most potent of these are checked perspiration, or unguarded exposures in variable climates, over-clothing as much as under-clothing of the skin, stooping posture, or confinement of chest by ligature or stays, the influence of absorbing passions, &c.; and most of all (in the highly-favored classes who should otherwise escape the disease) inflammations, which congest or consolidate portions of the pulmonary tissues, and the treatment of which, as *hitherto managed*, entails weeks of wearisome confinement to the sick room; too often in the olden time the poisoning of the system, and the ruining of the digestive organs, by the excessive use of drugs; bleeding, blistering, low diet, and depletants, together with the depression of the vital powers by every other anti-Hygienic influence.* We shall give for the present the apposite case of the monkys in the Zoological Gardens of London, not a great many years ago. An elegant room was built for them. Every attention was paid as respects the quantity and quality of food. But one thing was wanting---*ventilation was entirely neglected!* In short, the functions of the *skin and lungs* were ignored. The consequence was, *they all died of tubercular disease* within a short time.

In conclusion of this part of our subject, we believe it may be laid down as an irrefragable truth, viz., that no one with perfectly acting lungs and skin becomes tuberculous; or being tuberculous, long remains without the arrest of the ravages of the disease.

Lactic acid is one of the products of the decomposition of the tissues, *and finds its chief outlet by the skin.* When the cutaneous function is impaired---(*and this impairment, we contend, is an integral part of tubercular disease*)---the elimination of the lactic acid is attempted by other outlets, *chiefly by the bowels.* Hence the prevailing acidity of the intestinal canal in scrofula and phthisis remarked by all who have investigated that point. Hence the partial and temporary benefit of alkaline remedies in those diseases. This acidity of the *primæ viæ*, and the derangements of the alimentary canal associated with it, are most common in infants and children. Hence their greater tendency to manifest the mesenteric forms of scrofula.

We challenge refutation of this position, viz., that imperfect blood depuration (*i. e.*, defective play of the lungs and skin) *and not directly bad digestion, or faulty blood-making, is the primary source of the vitiation of the solids and fluids characteristic of scrofula and consumption.* A careful analysis of all the phenomena, and an extensive generalization of the best ascertained facts regarding the causation of these diseases, can lead the honest and dispassionate inquirer to no other conclusion. For our own part we have devoted many long years to this research. The solemn and unalterable conviction of our understanding we have now uttered---and fearlessly, as becomes a

*I shall illustrate all this another day in my medical histories of some distinguished victims of consumption.

truth-seeker. The foregoing observations, therefore, are a high plea, if they do not constitute an unanswerable argument, for the Turkish Bath to be established among us, co-extensively with the evils it is designed and fitted to grapple with.

Let us explain as briefly as possible the mischievous effects on nutrition of impairment of the functions of the lungs and skin---*i. e.*, of the want of adequate supplies of oxygen to combine with the carbonaceous waste of the body, and to effect its elimination from the system. This point of view will exalt the utility of the Turkish Bath more highly in our estimation than aught else. We have already shown that oxygen is the first want of the animal economy---a want of infinitely more importance than even food; inasmuch as the products of decomposition demand abstraction and exit, *momentarily* as they are formed. Now, as the food contains large supplies of this most indispensable element (oxygen), is it a very violent supposition, or a very improbable hypothesis, its reception failing by the *lungs and skin*, that the economy in its pressing want of oxygen *borrows from this source of supply*---albeit at too dear an interest? or, as Liebig would express it, *converts the elements of nutrition into elements of respiration!* What more likely resource, what more natural, what easier, what more at hand, than when the food is decomposed in the process of digestion, and its elements set free, that a portion of the oxygen of the fatty and albuminous matters should be abstracted to supplement the deficit of that introduced by the lungs and skin? In this way, a radical vitiation of the alimentary principles would be effected, thereby disabling them for perfect nutrition, precisely to the extent to which they had been robbed of oxygen. The tissues formed from this faulty material would, of course, be defective, or diseased, in a corresponding ratio. This deteriorated albumen *we know* presents in the case of tubercular subjects. It will not fibrillate like the albumen of healthy blood. It *assumes*, instead, a granular, amorphous form—a form unfit for the nutrition of the tissues. Chemistry will, perhaps, tell us one day what precise things have taken place in the atmoic constitution of this deteriorated albumen. Is it a very far fetched and unlikely conjecture, that it has parted with some atoms of its oxygen for indispensable depuration? in other words, to diminish the evils of an excess of uneliminated carbon in the system? Are we assuming too much in calling it *deoxydated albumen?* But we are fortunately not left in the same uncertainty as to the results to the oily principle of the loss of a portion of its oxygen. Chemistry even defines and gives a name to this deoxydated oil. It is is a cholesterine—a form utterly unfit for nutrition. It abounds in tubercle! This we should expect. The liver is the appointed organ for eliminating the excess of fatty matters in the system. Cholesterine is a constituent of bile. When in excess in the economy, we have of course fatty liver. Now this fatty liver is peculiarly and pre-eminently the lesion of consumptive subjects. The fat and oils of their diet go into the stomach sound.

Here we find them in a degraded shape; *i. e.*, largely divested of their oxygen. What greater proof could we have of the principle we seek to establish—viz., that oxygen failing by the skin and lungs, Nature, in her dire extremity, when perfectly non-plussed, robs the food of it---as it were, preferring that the machine be kept in play at any hazard and expense, rather than to come to a stand at once---that the patient die slowly and gradually rather than suddenly. Need we wonder that blood-globules made of such deoxydated materials are of low vital properties, and that in proportion as the system is compelled to use this faulty material a progressive deterioration of the whole solids and fluids of the body takes place---to an extent, in the long-run, utterly incompatible with the functions of life?

This which affords, for the first time, the true rationale of fatty liver, for the first time also yields the explanation, at once of the emaciation characteristic of tubercular disease, and also of the efficacy of cod-liver oil in checking that emaciation, and mitigating the symptoms. By virtue of disabled lungs and sluggish skin, vitiated air, faulty posture of body, ligatures of waist, sedentary habits, close confinement in unwholesome chambers, breathing live-long nights an atmosphere unrenewed, and doubly tainted by the mephitic exhalations of bed, &c., &c.---oxygen having become an imperious want in the economy, not only in the food robbed of a quota of its oxygen, but the available fatty tissues of the body are laid under contribution. Nature has, in fact, deposited fat in its areolar beds for the purpose of supplying the necessary oxygen during seasons of inactivity of the respiratory organs and skin. Instance, point blank, hybernating animals, who commence the winter fat and awake lean! The same is the source of the waste in phthisis. Cod-liver oil, by presenting a large store-house of oxygenous supply, spares the adipose tissues, and so far is an invaluable nutrient element.*

*I answer two objections here: 1st. Why is the oxygen of the oil and albumen robbed, which have so small quantities of it to spare, compared with the starchy and saccharine principles of the food which abound in oxygen? I reply, that the oxygen is not readily get-at-able in the food in question, because they are hydrates of carbon—*i. e.*, combinations of water and carbon, which water would require to be decomposed before its oxygen was available. Now we have no proof that water is ever either formed or decomposed in the body. But the oxygen of the oil and albumen is more easily separated. Hence these principles suffer the robbery of it, and the consequent deterioration of their properties as nutrient principles. *Objection Second:* Is not your destination of food antagonistic to Liebeg's theory of heat-forming and blood-forming elements? I admit that it is; and I am prepared to prove, moreover, that Liebeg's theory of animal heat of the destination of food is open to fatal objections, which cannot be entered upon here. We give one example: If carbonaceous foods were solely or chiefly for respiratory purposes, what becomes of the highly carbonaceous rice and ghee diet of the Hindoo, living often in a temperature above that of his body? How is this carbon burnt off without burning him up? Liebeg's theory totally fails to explain these points. How is rice and ghee incapable of sustaining an Esquimaux? Suffice it, then, to state my convictions, that every chemical research instituted will only confirm my position that the oxygen of the fat of tubercular patients is appropriated as I al-

Finally, on this branch of our subject. No undemonstrable or as yet undemonstrated truth is clearer (to my own mind at least) than this, viz., that the available oxygen of the food is converted into an element of respiration or depuration whenever sufficient oxygen for the purpose is not forthcoming by the inlet of the lungs and skin, or sufficient carbon not eliminated by the same outlets. Here, then, is a grand impairer of nutrition---a new, and yet very old factor of disease, introduced to the notice of the profession! Is this not tracking to his lair a fell destroyer of the human race, who has long lain in ambush?

The practical views now suggested in connection with the Turkish Bath, when pushed to their legitimate consequences, will operate, we believe, a great revolution one day in medical treatment; and will influence for good the destinies of thousands of unborn generations! I challenge my respected medical brethren to refute the distinct proposition I lay down on this head, viz., deficient oxydation of the waste of the body lies at the foundation of most diseases---an evil aggravated in chronic disease, by the attempts of the system to compensate this defect by abstracting oxygen from the food!

Disprove this allegation who can.† Beyond all question, this infra-oxydation is the starting-point of gout, of rheumatism, of diabetes, of granular kidney, of fatty degeneration, of many forms of fever, and of some others of our gravest diseases. If so, what is pointed out as the cure of this state of matters? Less trust to mere drugs unquestionably; and more attention to open, and keep open, the body's safety-valves! This can always be done by the simplest natural agency. It would argue little wisdom and less benevolence in the All-wise and All-merciful Designer and Maker of all things, if we were obliged to go to the wilds of Peru for a remedy to a disease caught on the banks of the Thames, or in the meadows of the Severn! But fortunately for mortals, the "bane and antidote lie both before them." If I were asked to give a brief and distinctive definition of the Turkish Bath, I would say, It is that which claims the exclusive or pre-eminent power of physiologically opening the safety-valves of the living mechanism; or, in other words, developing a high activity of the depurating economy of the animal body; and so fulfilling the first grand indication for the cure of all diseases. If wielded by courageous and skilled hands, no artificial or medicinal system will be able to compete with it, either as respects the quantity or quality of its cures. How precisely adapted it is to arrest the

lege it to be: that supplementarily the oil and albumen of the food of such patients are laid under contribution—are deoxydated for depurating purposes, in the defect of the perfect duty of the lungs and skin.

†My "party"—the "party" of the Turkish Bath—will doubtless challenge the profession to this disproof—to invalidate or substantiate my position. So important is the question practically, so much will the truth of this view advance the cause of the Turkish Bath, that I entertain strong hope that some rich partisan will make it the subject of a Prize Essay for German, French, or British chemists to decide.

ravages of scrofula and consumption, all theory now declares—if facts failed to speak. And we do anticipate and predict an immense decline in the prevalence and mortality of these maladies from the time of the general establishment and patronage of the Turkish Bath among the Western peoples—now their greatest victims. Among the Eastern nations who use the bath this desolator of European hearths is almost unknown! *Ecce omen.*

CHAPTER II.

THE AGENTS OF DEPURATION; OR, THE EXCRETORY APPARATUS OF THE BODY.

THE whole body may be considered, in one point of view, as a grand excretory apparatus. The Lungs, the Skin, the Liver, the Kidneys, and the Bowels are but the more prominent organs for the elimination and outlet of the superfluous, wasted, or noxious materials of the system. The first three only of these constitute the subject matter of the present exposition—giving simply so much of their anatomy and physiology as is necessary to the explanation of their functions. We begin with

1st. *The Lungs.* On this function all that is relevant or demanded for our popular treatise may be comprised in a very few lines; and the briefer the more desirable, because we have much to say on the Skin and Liver—organs much more under our control, and, therefore, more subject to abuse.

The largest product of the waste or transformation of the structures of the body is *carbon.* This is indicated by the dark color of the blood returning from the rounds of the circulation—exhausted, devitalized, and loaded with the impurities of the body's decomposition, as well as with much of the refuse of the materials of recomposition, chiefly carbonaceous. The Divine Architect of our frames has taken corresponding precautions for its excretion or throwing out. The apparatus provided to this end is at once the simplest and the most comprehensive. The exclusive requisite is a membrane that shall admit the *diffusion of gases;* in other words, that shall expose the blood to the influence of the atmospheric air. This is all that is necessary to the outlet of the most poisonous elements of decay, and to the entrance of the supreme principles of vitalization. To purify is thus synonymous with to vivify. The air-cells of the lungs and the pores of the skin are, respectively, the great contrivances for this purpose. It is the function of the lungs and of the skin to fulfil this conjoint office. Aeration of the blood is thus the first essential of life. Remove a fish from the water, and the gill-plates—its lungs—dry and cohere. Aeration of the blood is impossible. The fish necessarily dies. In the earth-worm, leech, and other animals far down in the scale, there is nothing of the strict nature of lungs and gills; but other equally efficient means (for them) of aerating the blood are adopted. The change from venous to arterial blood is effected in small sacs, or vesicles, usually placed in pairs along the back, and opening upon the surface of the body by means of pores in the skin, called *spiracula*, *i. e.*, breathing tubes. Close these spiracles, and you as effectually kill the animal, as by drying the gills you kill a fish, or, by obstructing a man's windpipe, you "stop his vitals." In the earth-worm there are no fewer than 120 of these minute external openings between the segments of the body. In the leech they are only sixteen on each side.

Throughout the whole animal kingdom there is an intimate relation between the energy of the vital functions and the activity of the respiratory apparatus. In cold-blooded reptiles, as the frog, respiration is reduced to the very minimum; the vital functions are correspondingly low and languid. In insects, on the contrary, there is a large provision made for breathing. In them we find vital action excessive—even vehement. The common fly is reckoned to move its wings a thousand times in a second! Witness the activity of a hive of angry bees, of hungry or thrifty ants, and the large amount of heat they evolve! The quantity of oxygen they consume far exceeds, relatively to their size and weight, the proportion of any other living creature. In the animals at the other—the high—end of the scale, the blood is aerated by a minute capillary network of vessels spread on the walls of the pulmonary vesicles or cells. In man, it is calculated that 1,800 of these bladder-like dilatations are grouped around the extremity of each air-tube—making in all some six hundred millions. The larger of these tubes possess muscular fibres, are hence contractile, and therefore liable to spasms. Thus originates one form of asthma. The average amount of carbon given off from the lungs of an adult is about half-a-pound per diem.

The exhalation from the vast pulmonary surface is a far greater agent in the circulation of the blood through the lungs than the propulsive power of the heart. This is incontrovertible—and this fact alone speaks volumes in favor of the Turkish Bath.

2d. *The Skin.* It is a low, incorrect, and unworthy view of this grand organ to regard it only and simply as a protective covering to the body. It is in truth much more—a living, sensitive, breathing, exhaling, absorbing, excreting, eliminating membrane of exquisite structure and endowments. Herein many of the prime operations of life take place. The skin may truly be called a great appendage to the heart and lungs—being an equal co-worker with them in the circulation of the blood. It does for the larger or *systemic* capillary circulation what the lungs do for the smaller or *pulmonary* circulation. It not only rids the blood of its carbon and supplies it with oxygen, *but regulates its density*—evaporating its watery constituents. The skin is at once the grand drying, draining, and ventilating apparatus of the body. It is in itself an universally expanded lung, kidney, liver, heart, and bowels! It is the greatest medium of nervous and vascular expansion, and, therefore, the seat of thrilling sensibilities, and exquisite tactile endowments. Altogether, the skin is an admirable piece of Design—illustrating alike the Wisdom and the Goodness of the Supreme Architect. On the sound condition of this organ, as much as, if not more than that of any other, depends the comfortable working of the living machinery. Its sympathies are intimate and universal with every suffering member. On it are reflected *their* ailments; and *its* derangements, in turn, are sure materially to modify for the worse the play of the interior apparatus. Herein is apparent how potent, not to say how safe, a battery the skin presents for the reduction of disease. In fact, many acute maladies select the skin as it were the common sewer for the running off of morbid elements which have accumulated in the system; and which no over action of the bowels or kidneys by drugs has been of avail to eliminate. We speak of the *sweating-crisis* in fevers, for example.

The effect of leeches and blisters, and hydropathic fomentations and compresses, illustrates further the powerful sympathies of the surface with the textures and organs seated below. Everybody knows how in small pox, scarlet fever, and other eruptive diseases, *the battle is won or lost on the field of the skin*, according as its safety-valve-functions rise or fall. If the interior irritation can be safely transferred to, and retained on, the surface—all is well with the patient. Do we want a ready test of the state of health of any man, or woman, or child, yea, even of our horses or oxen? We narrowly examine the skin! Its hues and its gloss, its roughness or its wrinkles, its sallowness or its pimples, speak a language the wise and experienced well comprehend.

The skin is the greatest excernent organ—the principal outlet of the body. It is a complete web of nerves and blood vessels; its thickly studded pores constitute the vastest system of corporeal drainage. Four times more matter is carried out of the body by the cutaneous surface every day, than by the alimentary canal. Costiveness or constipation of the skin, *i. e.*, constriction of its pores—a locked-up state of its exudations or exhalations—is, therefore, a much more serious affair than the same condition of the bowels. The latter may be "bound" with tolerable impunity for a week. A few hours arrested function in the case of the former may produce the most deadly symptoms; and if it were possible to seal up all the pores of the skin *at once*, as by an impermeable varnish, the individual would die in a few minutes! This accident nearly happened to a famous pugilist some time ago at the Royal Academy, where it was sought to take a cast of him *en masse*. We can now easily explain the sudden death of the boy who, at the rejoicings on the accession of Leo X. to the papal chair, *was gilt all over*, to impersonate the age of gold!

The skin and the mucous membranes, or the inner and outer linings of the body, may be called and considered almost identical structures. Their functions are reciprocal—indeed substitutionary and convertible. Hence the intimate alliance for weal or for woe—the profound sympathies—existing between them, and their sensitiveness to take on and resent each other's ails and aches. They are the great highways of traffic with the world without, and the vital domain within. Through them must pass *in* all the elements of corporeal reconstruction—the vivifying atmosphere and electricity—the pure *ether* of God's firmament around us—the nutrient elements, or food and drink, with salts, alkalies, earths, metals, &c. Through the same membranes pass *out* the corporeal sewage, *debris*, or waste—all that has served the purpose of the animal economy. The obstructed functions of one or the other of these inner and outer investments of the body, originates the largest number of Acute Diseases; as in their permanent derangement lies the greatest source of inveterate Chronic Ailments. If we want thoroughly to purify the blood, permanently to increase the temperature, to enhance the reactive powers—to induce, in short, a radical renovation of the entire man, we must address ourselves *to exalt the functions of the skin!* In one grand point, however, these co-related organs differ—they borrow their chief nerves from different sources. These of the mucous membranes are *nerves of organic life*, and depend for their energy on the spinal *ganglia*, or centres of vegetative or automatic action. The sensitive nerves of the skin, on the contrary, belong to the domain of *animal life*, and derive their origin from the *cerebro-spinal* centres. But the organic nerves are here interspersed also for the purposes of nutrition, and for the absorbent and exhalant functions of the skin. These nervous connexions explain the exquisite morbid and healthy sensibilities of the skin and mucous membranes, as well as their intimate sympathies with each other, and with the centres of vitality—the brain and spinal marrow, the heart, the lungs, the viscera of the abdomen, &c. In this way all morbid impressions are transmitted from without inward. By the same mechanism, the cutaneous functions in their turn become deranged by sympathy with every internal irritation; according to the extent and intensity of the interior derangement, or visceral disorder, is the healthy action of the skin marred or prevented; becoming, in turn and reciprocally, a source of aggravation to the internal malady. All digestive derangements, for example, tell upon the skin; and conversely, all cutaneous disturbance *tells* upon the digestive organs.

The texture of the skin is divisible into three principal layers: 1st, the outer scarf-skin, or *epidermis*—a simple exudation and drying up of cells or scales, in a pavement fashion, pushed upward from the *dermis chorium*, or true skin, below. The scurf of the head is an illustration of the epidermic scales. It is a truly *excrementitious* membrane, and may not inaptly be deemed and designated a sort of *protection-varnish* to the vasculo-nervous web below. But as it is constantly generated, it is not a coating *intended long to be retained!* Like all the other structures it is of *cell* formation. Possessed of independent, inherent power of life and growth, each cell draws to itself the fluid residumen of the colorless part of the blood, and secretes a horny matter. These cells lie layer upon layer, constituting a sort of mosaic flooring. As the deeper layers are gradually pushed outward and become superficial, their fluid portion evaporates, and they are converted into dry, flat, extremely thin, and dense scales. The abnormal accumulation of these scales is seen in many cutaneous diseases. Now it is easy to conceive how a dense compact varnish of this sort, when accumulated beyond measure—when not periodically removed—when encrusted moreover with dirt—obstructs the vent of the pores; not even admitting the tiling or layers of scales to act as a valve, and rise with the pressure of fluid from below. In the same way it is apparent how, by soaking and scrubbing, we improve the *permeability* of the skin, and, therefore, increase its fitness both for exhalation and absorption. This horny surface-skin is principally dried albumen, with unctuous matters. Alkalies combine with these and constitute a soap, or detergent. Hence the universal use of a combination of alkalies with oil for washing purposes. 2d, the *dermis* or skin proper, or *chorium*, is an elastic network of fine fibres or strands firmly interwoven. In the meshes of these are enclosed little bags of fat—cushions you may truly term them—a regular padding, as it were, provided by the Supreme Architect, to enable the skin to resist the compressions and contusions it is daily exposed to, as well as to fill up any irregularities of the surface. These elastic cushions, with admirable foresight and benevolence, are made to abound in the soles of the feet and palms of the hands! 3d, between the upper surface of the true skin and the scarf-skin, is a separate and distinct layer of blood vessels, and nerves heaved up into little conical eminences, like tufts, or the pile of plush. They are called *papillæ*. Hence this fine sensitive nervo-vascular web is called the papillary layer. The color of the skin depends on the quality and quantity of the blood in these vessels. The circulation of those of the head, face, and neck, is much under the control of the nervous system, as is manifested in the opposite effects of fear or shame. The retention of the blood in these little vessels gives the mottled livid hue of the skin when chilled, and what is familiarly known as *goose-skin* —the *gorged tufts!*

Inflammation of the skin consists in persistent gorging and retardation of the blood in these *papillæ*. The *pores* of the skin are minute tubes about a quarter-of-an-inch long, and of a spiral course. A coil of this tube constitutes the perspiratory gland. On the lines of the palms of the hands and soles of the feet these pores present visible dots, 3,000 to the square inch, equivalent to seventy feet of drainage-pipe on every square inch of the body. If all the pores were joined end to end they would form a tube twenty-eight miles long! Conceive, then, the results of checked perspiration—but a few miles of this sewage-way blocked up. Yet such obstruction is more or less the characteristic of most chronic and acute diseases. In these cases the excreting functions of the skin are more or less at fault. It is either scurfy, dry, and burning, as in certain fevers and inflammations; or it is pale and dead, and parchment-like, as in long-standing digestive derangements. To compensate this interrupted function of the skin, the liver, the lungs, the kidneys, or the bowels, assume often a vicarious or supplementary activity—a sort of double safety-valve work. Under this double duty they are very apt to break down—being then unfitted either for their own or their supernumerary functions. Hence the gravest diseases are engendered. Here drugs are but too often a powerless resource, because a fund of life, hard to replenish, has been drawn upon, which only the *organic energies*, by repose, and diet, and regimen, bathing, and perspiration, &c.—all judiciously handled—can gradually restore. Hence the virtue of the Turkish Bath.

The amount of visible perspiration, as every one knows, varies with the exertion undergone, and the heat of the weather. The insensible perspiration, however, or the vapor exhaled from the skin, is a more uniform quantity—averaging from two to two-and-a-quarter pounds per diem.*

From all this showing, then, of the nature and functions of the skin, it will at once appear how pre-eminently fitted it is, if not intended, to be the battle-ground of the physician in his conflict with disease: 1st, from its being the seat of thrilling sensibilities—as, in a sort, an electric surface—it is the great medium of transmitting soothing or stimulating impressions to the brain and spinal cord, on the one hand, and to the viscera of the chest and abdomen on the other. The nerves may well be compared to a system of infinite connecting wires, or telegraph lines, along which intimations of every kind are transmitted to the extremities, and all intermediate parts, and back again from the extremities, &c., to the centres of

* The skin abounds in oil glands and tubes, analagous to the perspiratory. The unctuous secretion takes place most manifestly on the shoulders, on the face and nose, along the ridge of the eyelids, in the ear passages, and the roots of the hairy scalp. This oily product is sometimes arrested in its minute secretory tubes when the skin is either torpid or inflamed. The contents become solidified and impacted in the tubes. The projecting points get blackened with dirt or dust. When the tubé is forcibly emptied an animalcule resembling a wood-louse is found embedded in the little worm-like mould of the tube. The disease is called *acne;* vulgarly, "grog-blossoms." The uses of this oily matter are evidently to lubricate the skin, to impede its too rapid evaporation, to neutralize the soaking, relaxing effect of moisture, and to protect it against acrid substances. In the eyelids it evidently serves the purposes of a gutter, or eaves, to confine the tears and moisture of the eye. It keeps the cartilaginous cavities of the nose soft, and, with the hairs, serves to repel the intrusion of insects.

power. 2d, from its immense superficies—constituting it the largest drain or waste pipe of the body. 3d, and lastly, from its being an organ both everywhere patent to observation, and capable, without injury, of standing a little rough treatment when necessary.

But a still more interesting point of view of the functions of the *skin* than even anything embraced in these comprehensive details remains now to be developed. *Depuration,* of which (as we have seen) it is a principal organ, is very grand work, and takes the precedence even of nutrition in the rank of importance to life. But the highest, the first, the most indispensable function of animals, the skin shares in comnon with the heart and lungs. It justly boasts to be a coadjutor with them in the prime faculty of *circulating the blood. Without cutaneous exhalation there could be no motion of the fluids!* The vital current would come to an almost instantaneous stand. So that, however great our admiration may be of the economy of the skin, as the chief eliminator of the carbon and lactic acid of the system, our ideas of its supreme utility and importance will rise still higher, when we view it as an organ quite as essential as either the heart or lungs to the circulation of the blood. This is a point of view many are not prepared for. Nevertheless it is the truth. It is ground that, so far as we know, has not yet been occupied by the expounders of this "Oriental question;" and it is, moreover, ground that is decisive. On this alone the whole merits of the Turkish Bath may be safely based. Its partisans need seek no other. Herein alone rests its all-sufficient defense.

Some of the facts on which the true philosophy of the Turkish Bath is based may be easily comprehended, and very briefly summed up.

The blood, as is well understood, describes a two-fold circuit in the body. 1st, that through the lungs; 2d, that through the general system. The heart, a double organ, and as a great force-pump for each circle, is placed at the junction between the two. But, mark well, the propulsive power or force-pump function of the heart, extends only a comparatively small way in the route the blood has to travel, *i. e.*, only through the more capacious trunks and palpable vessels. When we come to the *capillary circulation* (which is by far the greater moiety of the whole) we find *supplementary local forces* invoked to aid the transit of the vital fluid. We say nothing here of the alleged influence of the ganglionic nerves—of the contractile power of the capillaries—of the affinities and reactions existing between the vessels and their contents. These may be good hypotheses, but they are not demonstrable agents. The grand motor power we have now to introduce, viz., *cutaneous and pulmonary transpiration, is* demonstrable and point blank. There is an exact analogy and co-relation between the functions of the leaf in plants and those of the skin and lungs of animals. [The lungs may be likened to an extended inward skin, rolled up into folds or convolutions, *honeycomb-wise*, for the purpose of close-packing.]

Now, the force or influence which promotes the ascent of the sap in plants—viz., the exhalation from the leaf—is one identical physical principle with that which determines the motion of the fluids of the body toward the exterior, viz., the transpiration from the skin and lungs. All liquids in connection with an evaporating membrane acquire motion toward that membrane. In other words, evaporation from living surfaces, or even from dead membranes in contact with liquids, causes the fluids to rise in the capillaries—thus producing motion or determination from behind, *i. e.*, from within toward the surface. The amount of motion is directly proportionate to the rapidity of evaporation; *i. e.*, stands in a fixed relation to the temperature and moisture, or dryness of the atmosphere. Capillary attraction fills the vessels, but it does not cause the fluids to rise. The motion of the fluids belongs to, or is derivable from, the evaporating surface. The immense transpiration constantly going on, in the state of health, from the large exhalant surface of the skin and lungs, produce a virtual vacuum within the capillary tubes whence the fluid or vapor is oozing. By the external pressure of the atmosphere, and in the case of the lungs, by the vacuum created at each expiration, the fluids are forced, or rather drawn, into the superficial vessels. In this way the blood acquires a decided movement and determination to the surface. This *vis ab extra* is no doubt aided by the other powers concerned in the circulation, as the contraction of the capillaries, the chemico-vital actions taking place in their extremities, &c., &c.

From all this it will be very apparent how the suppression of transpiration (as by improper exposure to chills and draughts when the skin is unfortified or bathed in sweat, or by states of the atmosphere in which moisture and heat or cold coincide, and, therefore, the conditions of evaporation fail) is followed, as a necessary consequence, by a check of this outward movement of the fluids. A primary essential of health, if not of life, is thus interfered with. If the power of vital resistance be not strong, or if, at the same time, the body be diseased and weakened, then occurs sanguineous arrest or stagnation—congestions of vital organs; and, in the same proportion, impairment of vital functions. The lay-reader will not marvel at the fatality of lung diseases, chronic or acute, when he reflects that the lungs are a great rolled up inner skin,* with tubes, like the branches and twigs of a tree, penetrating in all directions through that rolled up mass—a true congeries of cells to convey the air to its hidden surfaces and convolutions. Bronchitis coats over the lining of this branching air-tube with a viscid phlegm. Pneumonia

* Imagine a great net of the finest texture and material, some fifty yards of blond, for example, with a minute but very distinct bladder filling up each mesh, and all this rolled up into the size and shape of a sugar loaf; but from the apex or cone (the point) proceeds a tube, with dividing branches and twigs, precisely like those of a tree, penetrating the congeries of cells and blood vessels in all directions, to convey the air to its every convolution, and to its inmost recesses. This gives you a perfect idea—if a rough one—of the lungs.

solidifies the porous mass of cells which constitutes, as it were, the leaves of this imaginary tree. Apoplexy floods this whole structure with blood. Tubercle compacts and hardens the mass. It is a concretion in its effect equivalent to sealing up or obstructing the pores of the skin *with a close crop of warts!* In such a state of affairs how can transpiration take place? What becomes of the functions of the lungs thus beset? Imagine a large patch of these supposed warts ulcerating and bleeding, and coalescing into a seething crater of corruption, and the general disturbance and local desolation that will emanate from this morbid centre. There you have the essence of consumption—what may be called, after this figure of speech, the Etna or Vesuvius of the living man, rather say of the dying man! Even, without a figure, we talk of pulmonary *caverns.*

No fact, then, we think, can be established more clearly than this, viz., that whatever impedes exhalation from the cutaneous surface, or from the air-cells of the lungs, stagnates the circulation of the blood in the interior organs. If the stoppage of the exhalation be complete, the arrest of the circulation is entire and sudden. Death, with coldness and shivering, ensues. Hence we find that coating over a rabbit with pitch (by preventing exhalation, and, therefore, the circulation and oxygenation of the blood) rapidly diminishes its heat; in fact, asphyxiates it. The rabbit so treated dies in a shivering fit!

We have another beautiful illustration of this doctrine of suppressed transpiration in the phenomena of *epidemic cholera.* Whatever be the noxious agent or miasm that causes the disease, one thing is very certain, viz., that it operates to annul or paralyze at once both pulmonary and cutaneous exhalation. Hence the Turkish Bath, early had recourse to, would be the cure, *par excellence,* as it cuts short the cold stage of ague. The stifling old fashioned vapor and hot-air baths, under the bed-clothes, failed, because what was wanted was pure hot oxygen, and the lungs to have their due share of it. The essential of cholera is the draining away of the watery portion of the blood by the exhalant surface of the bowels! What remains is so much clot, or tar-like residuum that cannot circulate. The lungs are useless and the skin dead. Transpiration is abolished. Oxygenation is impossible. The living furnace won't draw! The carbon cannot be burnt off. Animal heat cannot be elaborated. Hence the deathly coldness and blue skin characterizing the disease, from the entirely venous nature of the contents of the vessels. When things have come to this pass, the vitality of the blood is reduced to the lowest ebb. Hence the simple chemical affinities gain the ascendant over the vital. The serum of the blood separates from the fibrine, and the channel of its outlet once being opened from the congested intestinal membrane (where the blood has retreated on being driven from the surface) there is nothing in the unaided powers of the constitution to stop the drain of vitality.

The salts of the serum, indeed, *operate as a cathartic* to each exhalant tube! The drain goes on so long as there is any serum to drain away. The primary conditions of life fail—the organic powers are brought to a stand. The system sinks, defeated in an unequal contest.*

But the evil of checked transpiration does not lie solely in the visceral congestions so produced; but there is, moreover, the arrest of the chemico-vital changes ever operating, both on the surface and in the interior of the body. Perspiration, for example, contains, as we have already remarked, lactic acid and the lactates of soda and ammonia—the products of the decay of the muscular tissues in which this acid abounds. During muscular exertion these products are largely evolved. Hence, if perspiration be checked under such circumstances by prolonged cold, or chill, then these decomposed materials are retained in the blood, or forced to be eliminated by the vicarious duty of other organs. This is the fountain and origin of rheumatism, gout, diseases of the kidneys and skin, erysipelas, fevers, inflammations, &c. Hence we see how the blood becomes doubly tainted, doubly charged with abnormal elements. The

* Yet even in this, the body's direst extremity, if the patient has not been already poisoned by the remedies, or if the constitution has not been impaired by excesses, or by chronic visceral irritation (as from drugging and dram-drinking), genial nature will usually come to the rescue. The vomiting and purging will stop from sheer exhaustion—from there being no more serum to drain away. The very collapse that follows gives the organism time to rally—to collect her forces for a final struggle with the enemy. In the calm that follows, the soft tissues constituting the greater part of the body, yield up the fluids that yet saturate them; and the salutary thirst created, brings fresh supplies. The vessels receive the new tribute, and contract down upon their diminished contents, and so the circulation once more recommences. The reaction is apt to be excessive—a grand source of peril in the convalescence.

We cannot dismiss this allusion to the cholera-question, without bearing an honest, but fearless and emphatic, testimony to the merits of Dr. William Stevens, the discoverer of the only true antidote yet found to the ravages of this fearful scourge of mankind. His *saline treatment* constitutes one of the finest illustrations of the application of the Baconian, or inductive method of philosophy, to disease and remedy, to be found in the whole range of medical science. In fact, the medication in question is perhaps the only instance of a *specific* the practice of the Art of Physic furnishes. Everywhere else we grope more or less in the dark as to the real *modus operandi* of medicines. But here, the precise ingredients that are drained away in the exuded serum of the blood are restored to it! The success of the treatment corresponds with the accuracy of the philosophical analysis that dictated it—only two or three per cent. of failures—while a host of rival modes of cure often lose one-half of the cases; sometimes three-fourths! But impartial historical truth compels us to confess that paltry professional jealousy and personal pique were long permitted to obscure this great discovery, and to rob the suffering public to a great extent of its benefits. In this he only resembles his great prototypes, Harvey and Jenner. Posterity will do him justice. Advanced now "in age and feebleness extreme"—his heart dead and his ear deaf to the voice of human applause, he may yet console himself that a grateful country will not quite let his memory die! It would have been, perhaps, sufficient for the glory of a lesser name, to have been among the first—if not the very first—of surgeons who planned and successfully executed the grand operation of *tying the internal iliac artery!*

oppressed excretory organs are far from being up to the mark of their own respective functions, let alone performing supernumerary duty. Digestion and assimilation are weakened in the same proportion. Herein is a new and independent source of the direct generation of morbid products. Thus is the *melee* of the suffering organism thickened, and confusion gets worse confounded.

3d. *The Liver.* As this is a great decarbonizing organ, supplementary to the skin and lungs, and one influenced powerfully by the Turkish Bath, its function falls necessarily for review in this place. Situated midway between the apparatus of supply and the organs of distribution, it acts as a reservoir of carbon and a *diverticulum* from the heart and lungs, straining off, before it reaches these organs, the surplusage of carbon brought by the *mesenteric veins* directly from the alimentary canal. But the liver does something more than rid the system, at first hand, of superfluous carbon. The bile is more than an excrementitious fluid. Before being ejected it is turned to account for the purposes of digestion. Thus is the liver wisely ordained to economize material, to subserve nutrition, even by refuse drainage-matter. It serves to sift and clarify the dissolved contents of the stomach and bowels. It checks the influx into the general system of excess of carbon coming directly from the sources of supply, and so takes the strain off of organs already sufficiently charged with the body's impurities. The *thoracic duct*, or great main-pipe of the lacteal system, carries the chyle (the newly absorbed nutrient principles) directly to the venous trunk terminating in the heart. But the otherwise disposable carbon is absorbed by the mesenteric veins, and so finds summary exit by the liver—multitudinous and complex ends accomplished by simple means that show wondrous design—mingled Wisdom and Goodness! The immense quantity of blood the liver receives from the coats of the intestines, and which it decarbonizes, places in a strong light the relief the due performance of its allotted work affords to its coadjutors, the skin and lungs. These three grand allies in the living economy intimately co-operate with each other, play into each other, substitute each other, sympathize with each other, suffer with each other, and have their diseases cured by the relief of each other. The failure of any one of this "triple alliance" imposes upon the others vicarious duty, *i. e.*, if they can do it; and where they cannot, disease is the consequence. The prevalence of liver complaints among the indolent, luxurious, and high-fed classes, and in Europeans living in hot climates after the dietetic fashion of cold countries, is not now difficult to account for. In the first place, their food abounds in rich carbonaceous compounds, the error being not less in quantity than quality. In the second place, the amount of stimulant liquors taken to propel along their heavy indigestible meals, aggravates the intestinal irritation by determining an undue amount of blood in the alimentary mucous membrane. In this case, the skin loses what the intestine gains; the sanguineous excess of the one causing its deficit in the other. In the third place, the want of adequate exercise of the limbs, lungs, and skin, fills up the measure of these evils. This it does by preventing that due waste of the body, that activity of the excernant functions which passes off with the least bane to the constitution, the superfluities of a full or pernicious diet, oxydizing and eliminating the impeded products of decomposition. Herein precisely lies the error people commit in hot weather at home, or in burning climates abroad. Herein is the philosophy of the bilious diseases then and there prevalent. Under a high temperature, the cutaneous functions require the most unimpeded scope, instead of being diminished or paralyzed by diversions of blood to the interior by congested mucous membranes, &c., all the effects of table excesses, of irritant food, drinks, or drugs. Hence the twofold source of the accumulation of carbon in the system. 1st, that in the liver directly, from a too heating, full, and fatty diet, especially in warm weather or in hot climates. 2d, that in the general circulation, or in congested viscera, from its impeded exit by the skin and lungs. In cold weather, on the other hand, or in cold climates, people are less bilious. The habits are necessarily much more active, to enable them to resist the cold. The limbs, lungs, and skin, are all in more vigorous play, and so effecting more completely corporeal waste, as well as throwing it out, burning up the fuel of the living furnace, exalting animal heat by quickened transformation of matter and the increased chemico-vital changes so brought about. To this extent, therefore, is the liver relieved of the supplementary duty it would otherwise be obliged to assume if the superficial outlets of carbon were locked up or acting under par. Hence, in cold weather, the comparative, if not complete, immunity from bilious disorders of persons of temperate and active habits. But in hot seasons or climates, there being little or no demand for carbonaceous diet as fuel to heat the body, the labor of its extra extrication must necessarily fall chiefly on the liver. Hence, this organ, taken aback by duty it is incompetent for, irritated and overtasked, falls into disorder. Nature often attempts to clear away the surplusage thus accumulated, in the shape of cholera, dysentery, diarrhœa, fevers, &c. The same explanation accounts for the popularity of such medicines as calomel, colchicum, dandelion, &c., that stimulate the functions of the liver and emulge its ducts. These intestinal irritants and disgorgers of loaded gall-bladder and bile tubes afford the needed relief, but it is only temporary. It is like borrowing cash, in the Palmer fashion, at 600 per cent.! But say only cent. per cent. interest, or fifty per cent., what follows? What must follow, but corporeal bankruptcy sooner or later! The spendthrift goes on for a time, leaning on the false prop that is to pierce and break him. Medicinal stimulants, like alcoholic, leave behind the necessity for their repetition in increased dose! And, note well, the stomach was never intended to be a depository of filth in any shape, and pharmaceutical filth is often the most abominable of all.

The stomach is only fitted, as designed, to receive the legitimate elements of the corporeal structures—the sound building materials of the body. Aught else is inappropriate, unassimilable, uncongenial; in fact, in a lesser or greater degree, acts as a poison, if it be not actually such. This is a principle that cannot be impugned. But this game of over-stimulating, over-helping, over-straining the liver, will not always continue. The day of reckoning comes at last. Long-enduring Nature gets into the *sulks;* she will endure and be "put upon" no longer. Functional derangement, under all this tampering and tinkering, ends in structural alteration. A prime organ of life gives way, profound general malaise and disorder follow in its train, and the whole fabric totters to its fall.

The biliary disturbances, whether periodical or continued, is the simple attempt to explode off the pent-up materials of disease; and, in sooth, what are most diseases but efforts of nature to rid the system of substances undrawn off by the excretories, by the outlets appointed to eliminate whatever is superfluous or injurious? In a state of the system so charged and ready for a morbid explosion, it is easily conceivable how little things may upset the nice balance of health, may drop a spark of fire, as it were, among combustibles; as, for example, an indigestible article of food, a convivial excess, mental worry, extreme heat or cold, &c. It is not so clearly apparent how the same cause, in one case, insinuates slow, lingering, but fatal disorder, and in another, carries off the patient by rapid cholera, inflammation, rheumatic, typhoid, or putrid fever, &c., &c.

A vast deal of low spirits, *ennui*, *tædium vitæ*, &c., of the easy and wealthy classes, arises sheerly from the deficient excretion of the body's waste, notably from accumulated carbon, from biliary impurities; the freest, best, and safest vent to which would be by the skin, as roused by the Turkish Bath. If these morbid accumulations were sudden, they would produce all the shock of a narcotic poison, sometimes immediate death or paralysis; but, accumulated piecemeal, the system gets time to accommodate itself to the poison, as it does with alcohol, or opium, or arsenic, in large doses, if gradually begun with and long persevered in. But this very tolerance on the part of the constitution is the cause of the digestive and biliary derangements of the over-fed and under-worked classes. With so palpable a *materies morbi* gorging the liver, floating in the circulation and poisoning its life-springs, its particles arrested, perhaps, in the delicate textures of the brain, is it any marvel that patients are consumed with all sorts of nondescript bodily aches and ails—worst of all, with mental misery, far more intolerable than corporeal suffering? "A peerage or a pension," as the *Times* would say to the physician who should successfully exorcise these demons of our high civilization, the plagues of our most refined society. In the Turkish Bath,* conjoined with diet and regimen, air, exercise, and discipline of the appetites and passions, lies the remedy.

* Well regulated, *bien entendue*, and not prescribed at random, or to be invoked

CHAPTER III.

THE MEANS AND APPLIANCES OF THE BATH — RATIONALE OF ITS PROCESSES.

It is a sound axiom, universally received and acted upon by philosopical physicians, viz., that the disordered organism, given fair play to, rights itself; rectifies its own derangements; and it is, therefore, a principle held by some of the great practitioners of our time, one ably contended for by the late Sir John Forbes, that the cure of disease may be legitimately sought for in the due use of Nature's pure elements (*i. e.*, in the appointed or physiological stimuli of the vital powers; in the judicious aiding, abetting, and sustentation of those powers in their self-conservative struggles) and not exclusively in the vain nostrums and farragos of the apothecary's art! These may be all good in their place. The alleged "specifics" are nonentities, are a fallacy, a delusion and a snare! We have no specifics. Science renounces the research. Not more nonsensical was the pursuit of the "Elixir Vitæ," the "Aurum Potabile," the "Philosopher's Stone." My Lord Palmerston would define to a T the function of the physician *as being "the judicious bottleholder" to Nature!* This is really, in a great crowd of cases, the grand part he has to act. Now, we shall see what salutary ingredients the Turkish Bath puts into this restorative bottle: how it relieves Nature of the impediments that shackle her operations, how it softens and relaxes the solids that the fluids may the more freely circulate, how it expands and opens up the vast porous structure of the tissues, and so promotes the clearance and cleansing of the secret rills, and channels, and reservoirs of life. It sensibly seeks to purify the vital currents by flushing the vital sewers! It opens up the waste-pipes of the body, only to run off and disgorge through them its accumulated filth. The pores of the skin constitute, in fact, the vastest drainage-system of the animal economy, and are at once the safest route and most salutary outlet for purging off all extraneous, decomposed, or superfluous matters. The Turkish Bath sets about this scavenger-work by the immersion of the body in pure hot air. A preliminary macerating, sweating, clarifying, and eliminating process is thus performed. The pores are again closed, and the relaxed tissues and skin contracted, tonified, and braced up by tepid, then cold ablutions. Renovated vigor is thus imparted to the whole organism, even without the refreshment of food! Thus a grand immediate benefit is gained by this truly artistic process, viz., to

at the beck or whim of every patient who has once experienced its solaces. I happen to know that already the Bath, like other good things, is being abused. Thus a good cause will, by and bye, get discredited.

nourish and strengthen the body upon the old materials existing in the storehouses of the fabric, to burn them off, or to use them up, so as in any case to have clear receptacles, and clear conduits, for the elaboration and distribution of the new food. In this way we notably energize or activate the absorbing powers, the threefold effect of which is: first, to promote perfect circulation; second, to break up and remove unhealthy tissues; and, third, to put down more substantial structures in their place.

It may be received as a companion proposition to the first we stated under the present head of our subject—perhaps almost as a corollary from it, viz., that all irritation by drugs, violent corrosive substances (or by concentrated alcholic stimulants), of the delicate internal lining of the alimentary canal is equivalent to blistering it! Give a strong healthy dog a dose of what is considered a "mild domestic" medicine—"grey powder," with castor oil, or salts and senna. Dissected the day after, the mucous membrane of the intestines will present, here and there, large blood-shod patches—telling how the blister has acted. And yet we every day so blister the gastric tubes of delicate infants and children—not to talk of the horse-blistering in the case of adults—by aloes and colocynth, calomel and drastic salts, scammony and gamboge, elaterium and tartar emetic, Croton oil, *et hoc genus omne.*

Now, the Turkish Bath is wholly antagonistic to this destructive stimulation of the most delicate, sensitive, and highly vitalized surfaces of the body; tissues "tender as the apple of the eye"—as repellant to rude touch—as resentful of abrasion—and as difficult to appease when irritated. But the Bath not only does not irritate, it positively soothes man's sentient inner and outer linings, at the very time that it opens, and flushes, and floods the body's natural drains.

As the internal organs, therefore, are nice things to tamper with, or rather won't safely bear tampering with, Nature sets before us the skin as the grand battle-field in the warfare with disease. The keeping of this field in proper trim is also the best means of preserving health regained or not yet forfeited. Everywhere else the system may be refractory to our operations, and impatient or irresponsive to discipline; but the skin is always placable, always submissive, ever ready to be soothed or coaxed; and failing that, is not unwilling to be coerced into salutary action for the rest of the economy; provided always we know the right way to evoke its powers and to conciliate its co-operation.

The most fertile sources of morbid elements in the blood are retained or altered secretions. These are now admitted to lie at the foundation of a great majority of diseases; hence the most theoretically-feasible, as well as the most practically-available, agents of cure, are those required for the healthy exercise of the natural functions, especially those of waste and repair—of secretion and excretion. In the capillaries chiefly, if not exclusively, are carried on these processes of waste and repair—the building up of the new fabrics, and the taking down and taking away of the old, worn out, or useless materials of the body. Now the principal—at least the most demonstrable—seat of action of the Turkish Bath is the capillary system; its grand effect is thoroughly to open and cleanse the capillary tubes and strainers—to clear out their obstructions, and freely to circulate the blood through them.

The chief help Nature requires in most diseases, chronic and acute, is—first, to open the safety-valves, to rid the body of its impurities; then to establish the equilibrium of the blood alike in the central and superficial parts of the body—to soothe the sentient external surface, and to allay internal irritation---to relieve laboring viscera of intropelled fluids (*i. e.*, of congestion or stagnation). This purifying process---this inward unloading of organs---this equable distribution of the blood—is the sure, if not necessary, result of active determination to the exhalant surfaces, and the powerful drain therefrom of fluids easily and promptly replaceable.

Now, these aims just specified are the curative aims and "indications" of all medical practice, no matter what outward badge the practitioner may wear—what sect he follows—what name he is called by. That which best accomplishes these aims must needs be the best curative agent. The Turkish Bath, we conceive, unquestionably makes good this pretension, and is, therefore, the agent that comes nearest to the beau ideal of curative art. Above all other systems of healing, it is *par excellence* the equalizer of the circulation—the unrivaled and unfailing derivative to the surface—the solvent of capillary engorgements—the dissipator of morbid accumulations—the opener up of the body's safety-valves—and the flusher of its common sewers and drains; in short, the clean-sweeper-out of all filth blocking up the life-channels and poisoning the life-springs.

These are the direct and immediate effects of the practice we advocate. The indirect and the remote effects are, the increased quantity and improved quality of the secretions, the regulation of nutrition, and, in a word, the exaltation of vitality in the whole organism. In this way alone can we rationally hope so to aid and sustain nature as that she will be able to throw off most of the diseases that assail the fabric.

How, then, does the Turkish Bath accomplish all these salutary effects? How does it establish claims to efficacy such as no drugs and no system of medicine can pretend to? All this we shall proceed now to explain.

The first essential element of the action of the Turkish Bath is hot air; the purer the atmospheric oxygen, and the freer of all admixture or dilution, clearly the better. Under this stimulus, the whole secretory activity of the system is roused, transpiration is powerfully increased, both from the skin and lungs, with the effect of imparting extra activity to the circulation—a point sufficiently established in describing the effects of exhalation from the surface of the leaf in plants. This sanguineous *molimen*, or determination, is not merely on the surface; but it is effected from within, and to the surface. Every vital, vegetative, or purely organic function is stirred up to unwonted activity; the heart beats with renewed energy, and the blood vessels participate in its augmented impulse. The skin at length opens apace, however bound, obstructed, or reluctant its outlets at first may have been. With the pouring forth of perspiration, and thereby the absorption or neutralization of an immense amount of the surplus or latent heat of the body, comes instantaneous relief---a subsidence of the whole physiological tumult, raised expressly, as it were, to drive out an intruder. The large demand for vital fluids set up on the surface, and the chemico-vital elaborations there taking place, tend powerfully to unlock, and draw away, the pent up blood of diseased interior structures, congested viscera, and the like. The " change of matter," or " the transformation of the tissues," over the whole body, is facilitated; in other words, the waste of the animal structures is largely augmented. This demands the quicker elimination of this waste. With the increased outpouring of the structural debris---veritable body-sewage---unhealthy elements imprisoned within, are loosened, set afloat, and swept off by this real flood-tide of fluids.* speeding onward to the surface, like rivers, to be lost and exhaled in the ocean. The completeness of the aeration of the blood corresponds in degree to the activity of exhalation; respiration is deepened, and the lungs are profoundly fillled.† These actions now described are the most powerfully alterative we know. The effect on nutrition, the correction of its aberrations, is not long to manifest itself.

All this profuse drain of liquids oozing out by every pore of the surface, and drawn from every depth and cranny of the interior, justifies and calls for proportionate supplies of water by way of drink. This new fluid in its turn is drained away—thus literally washing out the blood, dissolving and straining off its impurities, and scouring out even the vessels. Absorption therefore, is not less quickened than elimination. Renewal and waste thus run a race with recruited powers. No morbid humors or even hard deposits, can long stand this perturbative, or break-up process, provided only it be judiciously repeated, so as not to impair the strength, or exhaust the stamina of the subject. In this way, excessive fatty deposition is broken up,

**Suspended internal functions* of various sorts have thus a *chance* of being set free from fetters that may have long enthralled them: and with this vent given to pent-up nature, the bloom of youth is restored to many a pallid cheek, especially in the case of young females. The simple draining-off of the overabundant watery elements of the blood of the subjects in question is no mean service rendered to the constitution, and paves the way for the filling of the vessels with purer and healthier materials. Of course, to do these cases full justice, they should be under professional superintendence.

†Hence the beneficial effects that may be legitimately expected in chronic congestion, hepatization, tubercular deposits, &c., of the pulmonary organs. But as these are the nicest of all cases to treat, they require careful surveillance, as well as accurate diagnosis. No random dosing will do; otherwise debility, rather than strength, may soon result.

melted down, and swilled out of the system; gross morbid humors of various kinds, and unhealthy tissues, are absorbed and removed. The muscles are rendered more compact; the skin tenser, more elastic, more clear, more glossy, more satiny, as well as more permeable. The same activity of absorption which takes down the paunchy and the bloated, also promotes the fattening of the lean and ill-nourished; and this, not only because the nutrient materials in the stomach are turned to better account, but because their resorption into the circulation is more energetic.

We have made no reference here to the action of the Turkish Bath on the Great Sympathetic System of Nerves. The stimulus of heat must powerfully affect these nerves, as well as the ganglionic and common sensory nerves. In like manner acts the stimulus of *cold*, which is also an integral and essential part of the bath. The organic functions, or the purely vital and vegetative actions of the economy, are much under the influence of the grand sympathetic and ganglionic nerves; and, therefore, it is to be inferred that we could have no increase of circulation, exhalation, secretion, &c., without the stimulation of these nerves. It may be demonstrated another day that in this sympathetic and ganglionic stimulation lies the whole curative virtue of the Turkish Bath, inasmuch as it is the forerunner and exciting cause of the augmented physiological actions that constitute the peculiar phenomena of the Bath.

The shampooing process, if not an essential, is a usual accompaniment of the Turkish Bath. Skillfully and moderately performed, as befits the less pliable frames of the hardier nations of the West, it will necessarily receive due attention, especially wherever the grand object of the Bath is to substitute exercise. But the subject simply requires allusion to here, not elucidation. At the end of the above described macerating ordeal—when the muscles, blood-vessels, nerves, and skin are all relaxed—is the proper time for kneading the body, in the same way as iron is best moulded and welded, and fashioned, when hot---an apt simile of Mr. Urquhart's.

The bracing, fortifying discipline of tepid and cold ablutions properly succeeds to the preliminary procedure of stirring up the circulating system, softening the surface, opening the pores, and producing purgation and waste by the skin. After thus giving vent to effete matters, or retained excretions, this conclusion of the process, and closure of the pores, is a *sine qua non* of the Turkish Bath---following up and confirming its benefits. Without this finale, its efficacy would be impaired, if not forfeited or lost, for a great many subjects. The unreflecting, or the totally inexperienced, may shrink at the idea of this sudden transition from high temperature to a cold bath, as something dreadful to bear, or dangerous to practice. But it is neither the one nor the other. The fear is a fallacy; the apprehension entirely groundless. On the contrary, the application of cold after perspiration in this fashion (passive) is not only not dangerous, but it is highly salutary and refreshing---exhilarating, in truth, beyond any previous conception of the uninitiated.

This conclusive operation is based on the soundest physiology, and is not less needful and appropriate than it is grateful to the patient. A general maceration of the tissues has been effected. The vessels, and nerves, and skin have been all relaxed from the heat and stimulation they have been subjected to, and from the copious floods that have oozed through them. A virtual depletion has been effected, the only depletion that is sound and safe. Now, then, is demanded, and is borne, the shock---the bracing power of cold. By this the cerebro-spinal and ganglionic nerves have temporary excess of vitality at once imparted to them---a veritable electric thrill is felt. A rush of blood is determined to the surface, to replace the heat abstracted. The effect of this is to increase and fix the circulation in the skin, thus rousing the capillary actions of the surface at the expense of the interior; promoting thereby the dispersal of congestions, and establishing the sanguineous equilibrium of the central and superficial parts of the body. All this brings about a rapid "transformation of the tissues," the breaking up, absorption, and swilling out of old, decayed, or diseased matters, and the deposition of new. The normal, or physiological activity of the vital functions is increased, the *vis vitæ* exalted everywhere. The more freely the skin has been acting, the larger the flow of fluids, the greater will be the cold that is desired; the better will it be borne; the more potent will be the stimulus it affords; the more permanent the reaction that will ensue; the more decided, in short, its curative results. Hence the feeling of immense relief and solace, of renovated mental and corporeal vigor, after a process that, to the superficial thinker, seems exhausting.

The phenomena above described, are vaguely expressed by the word reaction. In this reaction itself lies a great aim and agency of cure. To be able to react well is the grand help nature requires in a majority of diseases. The body corporeal then does for itself, for its enemies within, what the body politic does for itself when it rises *en masse* to repel its enemies without. In both cases, the effect is at least to quell or appease internal irritations, dissensions, and tumults!

By the discipline of the Bath, any over-sensitiveness or morbid sensibility of the skin, becomes so blunted, its tissues are so braced and fortified, its natural functions so exalted, as to bear with impunity any transitions of temperature, and the more extreme, often the more agreeable; as also the more hardening the effect. With the restoration of a high condition of the skin, coincides the return of healthy functions in the mucous linings, whether of the lungs or of the alimentary canal. In this way persons that are subject, on slight exposure, to catarrh, influenza, bronchitis, diarrhœa, &c., get case-hardened to atmospheric variations, and even bear draughts with impunity.

The allegation, that perspiration is a weakening process, is another fallacy that hardly needs demolition. Sweating, as accomplished by drugs (sudorifics), we admit is a debilitating drain. So is the vapor bath as used in the bungling way common in our old bath establishments. But properly evoked, and followed by tepid, and then cold ablutions, it is, on the contrary, highly tonic and invigorating. In the Turkish Bath, the patient lies full-stretched, in perfect repose, on couch, bench, or *dureta.* Nothing of the normal constituents of the body is abstracted save the saline and watery portions of the blood. The water is replaced by absorption from the stomach as rapidly as it is given out; for when the drain comes to be excessive, the supply is proportionate. And here be it well observed, it is only in very pure systems that the water, welling out from the pores, comes away pure. It is far otherwise when the body is impure. Not only the water oozed out by the pores, but the atmosphere all around is tainted by the eliminated products and exhalations of disease. This happens in bad cases of chronic maladies, characterized by corrupt humors, constitutional taints, &c., *i.e.*, whenever the secreting and excreting functions are materially interfered with; whenever in short, substances are retained either in the highways or the byways of the circulation that should have been eliminated. These constitute a very formidable, as a very palpable and intelligible *materies morbi.* In granular kidney (Bright's Disease) these odors in the calidarium are occasionally something dreadful. The easy exit afforded to these pent-up elements of disease by the powerful drains and perturbative action of the Turkish Bath is, beyond all contradiction, the source of its immediate and permanent benefits. Hence, if skilfully wielded, the reputation it is likely to achieve in the cure of visceral congestions, morbid accumulations and obstructions, and in blood-taints, &c.

If the Bath fails, nothing else will avail to transfer to the robbed, emptied, shriveled, parchment-like surface of the body, blood long pent up in a torpid liver, an engorged spleen, a congested mucous membrane, or a hepatized lung. By its outlet of peccant matters it gives immediate relief to *malaise*, misery, and fatigue. Increased absorption and elimination remarkably improve the appetite, and promote digestion and nutrition; healthier solids and fluids are formed than those that are thrown out or wasted down. Hence the Turkish Bath fills up the skinny or flabby, and reduces the obese, the paunchy, and the plethoric.

CHAPTER IV.

WHAT THE ANGLO-TURKISH BATH SHOULD BE.

This is a very brief "head" of our subject, and may be very summarily discussed.

The "performance" of the Turkish Bath consists essentially of "four acts," requiring indispensably as many separate chambers; *i. e.*, for any good public establishment.

1st. The first of the suite is the dressing and cooling room. Here the bath costume is assumed, and here the bathers rest, cool, and dress at the conclusion. The size, style, and arrangements of this apartment admit of every variation; but it should, wherever practicable, be a large "hall" in dimensions. To modify the details conformably to English habits and tastes, it should have stalls screened off, with couch, chair, dressing table, &c., in each, in order to afford all the privacy and convenience desired. Spiracles, or ventilating "bull's eyes," should be placed at the head of each couch, to give the bather full command of his atmosphere, independently of the general aeration of the room. The dome form of ceiling would be the most suitable, as the most wholesome, certainly the most picturesque, if not also the most Oriental, in style. The gratification of taste and ideality is an almost indispensable condition of attaining the fullest benefits of the Bath. All the pocket-questions, however, of any proposed building, we must leave to those whom they concern.

2d. The second chamber—the Tepidarium—is of an average temperature of 115 deg. to 135 deg., with a wooden platform covering the flues all around the sides of the room. On this bench mattresses are laid, or couches are scattered about. Here the bather reclines, and perspiration commences. This process may last from twenty to forty minutes—longer or shorter, according to the habits of the individual, the tolerance of the constitution, or the necessities of treatment.

3d. The Calidarium, or Sudatorium of the Romans, is of an average temperature of from 140 deg. to 160 deg. This is high enough for all salutary or sanitary purposes, and even too much for a great many people. In this, the true "hot chamber," the perspiration becomes very profuse, and the operation of shampooing is performed, *i.e.*, the muscles and integuments are kneaded, and the joints stretched and twisted *secundum artem*. When "enough" has been had of this *multum-in parvo* exercise (and the chamber should be left before there is any symptom of fainting), the bather is supported by the bath attendant to

4th. The Lavatorium or Frigidarium—the Bath-room of the establishment. It should be replete with every convenience for tepid and cold bathing, douches, pumps, jets, &c. Here the patient is well soaped and lathered, washed first with warm water, or a warm rain-bath drenching him at all points. Then he finishes off with cold douche, warm bath, or plunge. A good deal of animal heat is abstracted in this chamber; and it is prudent to replace it by a return to the hot-room for a few minutes. From this the dressing-chamber is once more gained, where the bather, enveloped in fresh bath costume, seeks his couch, and cools gradually. Now the real elysium of the bath is enjoyed, without any risk of catching cold. After half an hour, or longer, of the most intense physical luxury and mental calm, the bather resumes his garments, and takes his leave—his predominant feeling being the enviable one of the traveler in the desert who comes to an oasis, a green, fertile, well-watered spot, amid the arid wilderness of sand he has traversed. The bather has enjoyed at least a two hours' oasis in the desert of this world's cares; and he therefore quits the establishment, not only a purer, but a devouter man, perhaps, than he entered; feeling, we hope, his heart raised in gratitude to the Giver of all Good, for at least one additional sweet mingled in the cup of life, and with a very sensible bracing-up anew for its battle; but feeling most immediately concerned to eat, and prepared to digest, a good meal.

Before proceeding to the last part of this little treatise, it behoves us briefly to allude to a much disputed practical question in connection with the bath, viz., how much, if any, moisture should qualify the atmosphere of the hot-room? This, in our humble opinion, is a question very easily settled; one, perhaps, which should never have raised any controversy. We have only time to touch on four points here, which, we hope, will be decisive. 1st. Let the reader recollect what we have said on the subject of evaporation, and the conditions promotive or repressive of exhalation from membranes. Now, in the bath, the skin is moistened with perspiration; but in proportion as the air is loaded with an excess of aqueous vapors, the second condition of transpiration fails—the less necessarily is the evaporation from the lungs and skin. Hence the blood will naturally tend to retain its aqueous constituents. They will, therefore, be in excess; the tissues will be inclined to paleness and flabbiness; in other words, to serous infiltration, or the leuco-phlegmatic habit of body. This we see exemplified in the pasty complexions and relaxed tissues of the inhabitants of the humid Low Countries.

2d. Vaporization is concerned in keeping down the temperature of the body—is a cooling process, and one of the active agents (as we have shown in its place) in the cutaneous and pulmonary circulation. When, this, therefore, is interfered with, or prevented, the blood is necessarily of a higher temperature. Perhaps, for the same reasons to a certain extent, but assuredly because of the deranged circulation following impeded exhalation, we have internal inflammations, determinations, and congestions favored.

3d. Evaporation is one of the steps of the blood-making process.

The superflous watery particles of the chyle are thus let off in its passage through the lungs and skin, and by this means its fibro-albuminous principle is advanced to the more perfect condition it presents in the *liquor sanguinis.* In the case of the Hollander just alluded to, not only may the watery constituents of his diet not be sufficiently exhaled, but his chyle may be marred in quality from the same deficient transpiration from the skin and lungs. Hence a double taint of his blood—the twofold origin of his constitutional characteristics.

4th. A moist atmosphere is a rarefied atmosphere, and thereby diminished in its respiratory qualities—the elastic distention of the dead vapor displacing a portion of the viable air.

On these clear and unequivocal grounds do we give the award to Dr. Barter, as right alike in theory and in practice—as having both science and fact on his side—in his views of the Turkish Bath; contending strenuously as he does for a calidarium, or hot-room, without any admixture of vapor!

Theoretically, then, the argument is all on the side of those who contend for the entire absence of vapor from the hot rooms of the Turkish Bath. But practically the question may be settled in its favor, provided the amount of vapor be so very small as merely to soften the aridity of a perfectly dry atmosphere.

CHAPTER V.

THE BATH CODE—RULES AND REGULATIONS FOR ITS SAFE AND SALUTARY ADMINISTRATION.

1st. Calm and repose of mind and body, is the first essential rule of conduct in the Bath. All distracting thoughts and passions, therefore, should be left at the door, or laid aside with one's garments. Even to talk is more or less to excite the brain, and should be avoided as much as possible. The reason of this rule is obvious, because the object sought is to summon into vigorous exercise the more organic or vegetative powers of the economy—to promote for the time the quickened activity of circulation, exhalation, excretion, absorption, &c., and to set at rest the jaded or worried animal nervous system; in other words, to quiet the brain, to soothe the sensitive nerves, and to rouse the organic or nutritive nerves.

2d. In cold weather, bathers going in with cold feet have the first cup of comfort, so to speak, administered in the shape of a warm foot bath, which can be taken at the bather's pleasure for as long as he pleases, and as often as he pleases. The feet being well warmed at the off-go in this manner, the determination of blood to the head some bathers complain of will be quite obviated, and perspiration will be induced in a very much shorter time.

3d. If the uninitiated bather feels faintness, palpitation, or difficulty of breathing, he should either go back to his dressing couch, and open his ventilator, and so get "a mouthful of fresh air," or he should have his feet and legs soused with cold water, and sip some cold water, or have a compress (or small towel), wrung slightly out of cold water, and applied to the region of the heart or stomach pit, with a tumbler of cold water dashed over the face, or slightly tepid water, to cool the head and produce evaporation; or a wet compress may with advantage be rolled round the head, or he should conclude the process at once by the cold douche. But to bathe the face and head only is sufficient, and the safest. But let bathers beware of cooling the head too freely, even with tepid water. Cold water for that purpose is totally inadmissible, because of the undue reaction to the head that would infallibly result. But even the tepid water cooling of the scalp must be cautiously and rarely done; once should suffice. If "headachy" feeling prompts to repeat this head cooling, be sure that the continuance of the bath for that occasion had better be dispensed with, and hasten to the lavatorium, and take a tepid rain-bath, which is ordinarily sufficient for almost any case. By any of these expedients the commencing faintness, the palpitations, or the breathlessness, will be at once arrested; and the bather may thereby be enabled to continue a sufficient time in the hot-room to flush the vital sewers, to open well the perspiratory outlets, and to bring relief to internal irritation, inflammation, or congestion.

4th. If the object of the patient is to perspire freely, and the sweat is reluctant to break forth, the tepid ablution of the head will enable the patient to stand a larger dose of the heat; but he must be very cautious in this procedure, for the after result would be an undue congestion of the face, with headache and evident fulness of blood. The time spent in either the hot or tepid room will vary according to the sensations and power of tolerance of the bather. A very susceptible subject should suffice (at first, at least) with the tepid chamber; its temperature is high enough for ordinary salutary purposes. From half an hour to an hour is a good average for most people. The profuseness of the perspiration, of course, is the grand modifying element in estimating the time to be spent in the room; the *skin-bound* require longer time than those with whom perspiration is early and copious; but the former cannot often stand it so well.

5th. After the skin has been thoroughly macerated with perspiration, and the muscles relaxed by heat, is the proper time for the shampooing process. The denser tissues and more compact fabric of Britons won't stand this pummeling, this kneading and disjointing operation, like that of the lithe and relaxed Oriental. But as much as can be borne of deep handling, chafing, and *succussion* of the flesh, should be performed on every bather. The friction of the surface, the soaping and peeling off of the scarf-skin, follows this. Last of all comes the cleansing off with tepid water; and then, with robust subjects, the cold douche, wave bath, or cold plunge, should be used. The due amount of cooling required in the lavatorium for individual cases is a very nice point, as upon its due regulation depends the grand efficacy of the bath and its perfect comfort, as well as perfect safety. Ordinarily, the tepid rain-bath in cold weather is sufficient; only the young and the robust should take the cold douche or cold rain-bath, unless in very hot weather. Another cautional rule of practice is, never by any means to allow the cold douche or wave bath to hit the head. For delicate ladies and children especially, I quite dissuade any violent shocks in the way of sousing, spraying, or plunging in cold water. Many infants are subjected to very cruel operations in the cold tub every morning, by way of hardening them; and with children of really feeble constitutions the result is the opposite of hardening. This *finale* of the bath closes the pores, and braces the relaxed muscles, vessels, nerves, and skin; thus fortifying the whole system, and rendering the taking of cold impossible. A single minute of this effectual finisher-off suffices. It is sure to obviate any exhaustion or faintness felt in the hot-room. A momentary chill by the sensitive, or the weak, or the weary, may be felt. This is removed by the return to the hot-room for but a few minutes, or simply passing through it may suffice to replace the abstracted heat.

6th. The skin is best allowed to dry in a warm sheet, the patient taking his rest on a couch in his dressing-stall. Here he can denude himself without let or stint, if he wishes to free his breathing surface from all impediments—a most salutary and hardening habit, not to speak of the advantages of a much greater amount of oxygen absorbed by the nude skin than when covered. The time spent in the cooling-room—the luxurious time of the bath—may be about half an hour at an average.

7th. The bather should dress leisurely, and should walk away at his leisure, taking care not to renew the perspiration, for he may thus, if under unfavorable circumstances, expose himself to chills.

8th. In persons whom we may truly term hide-bound, in those of very callous skins, or those who have seldom or never perspired, every thing would be gained by soaking them in a bath of 98 deg. for ten minutes before commencing operations. A moist membrane is an exhaling membrane; and this would best secure a breathing condition of the skin, relax the pores, unbind the tense dermis, as well as soften and soak off much superflous epidermis. From half an hour to a full hour would often thus be gained in persons hard to perspire.

9th. The bath should not be taken on a full stomach, nor yet on one completely empty (*i.e.*, after a prolonged fast). One hour after breakfast or lunch is the best time. After a full evening dinner or supper, two hours should intervene.

10th. Water may be drunk *ad libitum* by the thirsty and the fainting, and moderately as a general rule.

11th. When the object is to deplete or disgorge congested solid organs (as the liver, the spleen, the kidneys), or to reduce a solidified lung, then profuse perspiration, at the sole expense of the existing fluids in the body, will be more likely to drain off the excess of blood from the overloaded organ. To this end withhold all drink during the process. But diseases of the kidneys are an exception to this rule. The matters strained off by these organs require to be well diluted in order to be washed out of the system by the perspiratory tubes and the exhalent outlets of the skin. In such cases, also, the terrible odor that is diffused through the hot rooms sufficiently indicates the unusual and deadly stuff which the lungs and skin are set to work to throw out.

12th. Lastly, moderate exercise, short of fatigue, preceding the bath, will increase its efficacy—chiefly, as materially shortening the time necessary fully to rouse the functions of the skin. In what I call "the hide-bound" this preliminary exercise is of vast importance.

CHAPTER VI.

THE LEGITIMATE MEDICAL DOMAIN OF THE TURKISH BATH—ITS PRACTICAL APPLICATIONS.

I. The Turkish Bath is the truest and best anti-spasmodic. In cramps of all degrees; in spasams of the muscles of the bowels, which are the source of the pains called colic; in spasms of the gall-bladder, and gall-ducts; in pains in the region of the kidneys, or lumbago; in spasms of the bronchial tubes (asthma); even in lock-jaw and tetanus, its use is a legitimate and hopeful experiment at least. Between the combined effects of the hot room and cold douche, spasms of any sort will have a better chance of yielding than under any other mode of treatment; but very hot fomentations with flannel must be conjoined. In any case, the tedious convalescence, the usual result of the powerful medicines swallowed to overcome spasams, will be saved.

II. The Bath presents a valuable resource in the reduction of dislocations, and of strangulated hernial tumors (ruptures).

III. The Bath will be of the greatest utility in passive diseased states, wherever action is below par, as in the very commencement of acute diseases, in the premonitory stage of fevers and inflammations—the stage of depression of power—in the congestive stages of eruptive diseases (measles, scarlet fever, small pox, &c.) wherever, in short, collapse takes place and the symptoms show retrocession of the fluids from the surfaee to the interior; in other words, wherever congestion of vital organs exists or is apprehended.

IV. The Turkish Bath, for this reason, is an unquestionable resource in cholera—will be, perhaps, its grand remedy in the first stage. Having already spoken at large of this disease, as likely to be influenced by the Turkish Bath, we need not enlarge here.

V. The Turkish Bath should be at once had resource to in the collapse, shivering, uneasy feelings and depressed spirits that follow a decided chill of the surface, when perspiring freely; as, for example, when getting wet in an exhausting journey, or from too long exposure in an open boat,* or from the absorption into the lungs of an infectious miasm—a dose of which a man often gets in standing over an open drain. In all these cases, before active irritation or acute inflammatory symptoms have manifested themselves, there is every reason to hope that many diseases would be strangled (to use the favorite phrase of French practitioners) at the very off-go, and thus many premature deaths, often of the most illustrious personages, would be prevented. Thus died the Duke of Kent! Thus died George Washington! Thus died Count Mirabeau!

VI. In purely nervous irritations of the heart, or in those connected with organic disease, in simple palpitations; in *angina pectoris*, the hot room actually does quiet the circulation, and would do so still more remarkably, we think, if the cold or hot compress, according to circumstances, were kept on the chest, and often refreshed.

VII. In the case of local spasms, hot flannel fomentations applied to the seats of suffering while in the tepidarium would probably facilitate their solution.

VIII. The Turkish Bath will diminish the liability to take infectious diseases. This often depends upon a habitually sluggish condition of the kidneys, with marked and scant secretion. The powerful revulsion to the surface, and drain of fluids by the skin, operated by the Bath, effectually takes the strain off the kidneys—disgorging them, and, in fact, almost performing their functions!

IX. In "Bright's Disease," in diabetes, in gout and rheumatism, and in all kidney diseases, with excess of uric acid and its salts, the practice that carries off the corporeal debris by the skin—and not by irritant drugs acting on the kidneys or the bowels—is the true art and science of their cure. In such cases water-drinking during the bath is strenuously to be insisted upon, inasmuch as the excess of water washes out a corresponding proportion of solid constituents. Thus colchicum, or acetate or nitrate of potash, may be superseded.

X. In Ague, the Turkish Bath offers the most feasible remedy, as being a disease resulting from diminished secretion of the solids

*Lord Byron thus caught the fatal rheumatic fever that carried him off. He had got wet through in riding, and, contrary to all remonstrance, would return home in his wet clothes and exposed in an open boat. The Turkish Bath, on his arrival at home, would have been the surest means to have counteracted such a chill. But, alas, what treatment had this great man in his last illness?

strained off by the kidneys. The probability is, therefore, that a highly active state of the cutaneous functions would eliminate these solid matters of the urine through the surface, even as we find an eczematous eruption occasionally frosted over with crystals of urate of soda.

XI. In tic-doloureux or neuralgia, the Bath promises great things.

XII. Skin diseases will most probably be removed by a very summary process in the Bath, according to all experience hitherto.

XIII. In irritative congestions of the windpipe, from public speaking ("preachers' throat," so-called), the Turkish Bath can hardly fail to be pre-eminently successful; for this disease is usually only symptomatic of a morbid condition of the skin and digestive organs.

XIV. In acute affections of the throat and tonsils, even in croup and diphtheria, the Bath will almost invariably save life.

XV. In consumption, the Turkish Bath, fairly tested, will, on the clearest abstract grounds, as well as on the showing of facts, produce the greatest ratio of arrests of the disease. The noxious acids of the alimentary canal are thereby drained out of the system, the air cells of the lungs are dilated, pulmonary secretions are dried up, internal congestions are dissolved and dissipated, the relaxed skin braced, appetite promoted, night perspirations checked, the noxious chills and shivering at once cut short, and refreshing sleep procured.

XVI. In digestive derangements characterized by intense acidity, the Turkish Bath offers a great resource, as oozing out through the skin the excess of latic acid, which often lies at the root of the evils of dyspepsia.

XVII. In chronic bronchitis, and emphysema of the lungs, and in the dry catarrh of the aged, the Turkish Bath is worthy of extensive trial.

XVIII. In dropsies, both of the shut cavities of the bowels and chest and of the exterior tissues, as well as from diseased kidneys, the Turkish Bath is precisely suited, and will work wonders—as taking the tension off the veins—the effusion of water being only a vicarious effort to relieve the plethora of the congested vessels.

XIX. In tympanitis and other cases of abnormal secretion of gas in the stomach and intestines, the Bath will promote the extrication of the gaseous exhalation, or suppress directly its formation.

XX. In chronic liver disease, in enlargement of the liver, and jaundice, &c., the Turkish Bath will be found the most potent agent of cure, as demonstrated by the large success of the much inferior hydropathic instruments of sweating used in such cases.

XXI. In gout and rheumatism the Bath will prove itself the speediest and best remedy.

XXII. In syphilis and mercurial diseases. In diseases arising from the abuse of treatment, the same hydropathic experience calls for an extensive use of the Turkish Bath. The medicated vapor-baths of the *Hopital de Mudi*, in Paris, are less efficient attempts in the direction of the Turkish Bath.

XXIII. In the large and too common and distressing class of uterine diseases, the Turkish Bath will supersede, to a very large extent, the often very tedious and (to the constitution) expensive medication, by means of caustic and the knife, mechanical helps, &c.

XXIV. Cancer has now, perhaps, found its antidote in the Turkish Bath. Mr. Urquhart communicates a remarkable case of a lady who came to him in a desperate and hopeless condition, after the cancer had once been excised, and who was so far recovered as to be able to walk five miles. We hope the profession will give a fair trial to this remedy in a disease wherein they admit the powerlessness of all ordinary agency.

XXV. The Turkish Bath will take down, summarily and safely, excessive obesity—literally melting down and oozing out the oil of over-abundant adipose tissues; draining, as it were, the muscular fibres of this paralyzing accompaniment, as well as thereby increasing the tone and motor-power of these fibres. The Bath promotes the nutrition of the ill-nourished, increasing the appetite in proportion as it increases absorption.

XXVI. In diarrhœa, dysentery, &c., the Bath will be the cure *par excellence;* as determining excessive action, and diversion of the fluids from the intestinal lining to the skin, as well as soothing ganglionic irritation.

XXVII. We are inclined to hope that the Turkish Bath will prove itself the nearest thing to a specific for hydrophobia. If anything will ooze out, or neutralize the virus, once perfectly developed, it will be the action of the highest temperature that can be borne. Last century it was the custom, in some parts of Scotland, to smother these unhappy victims, by placing one feather bed upon another, the patient between, and a party of women sitting all around on the edges of the bed. On one of these occasions, within the memory of a living individual, a little boy was put in to be so strangled. After a quarter of an hour, when they thought he was dead, to the surprise of the operators, in taking off the upper bed, he leaped up out of a pool of perspiration in the center of the bed, where he lay, and said he felt quite well—indeed he was cured! This is an encouraging fact for the trial of the Turkish Bath.

XXVIII. The Turkish Bath will undoubtedly prove itself the best corrector of what has been designated the civic cachexia, the vitiated habit of body bred by hard town-life, whether it be the life of luxury or the life of labor—a nameless, nondescript condition of the solids and fluids—impairing much, if not quite, the relish of life, rendering vapid its enjoyments; and all this the result of over-excited brain, over-worked stomach, over-gorged vessels, and under-worked limbs, lungs, and skin—the effect, in a word, of closed safety-valves.

XXIX. The Turkish Bath will become an indispensable substitute for exercise to three large classes of people; 1st, to the indolent and luxurious, who take advantage of their privilege, but who find it, alas! anything but a blessing to be exempt from the primal curse.

2d, to the brain-toiling, city-pent masses, the keepers at home, the men of literature and science, the drudges of the desk, the prisoners of the counter, or the slaves of the factory. 3d, to valetudinarian multitudes, not ill enough to be loosened from the oars of business—"which thousands, once chained to, quit no more"—but too ill for personal comfort, and for the comfort likewise of those around them—the hypochondriac, the bilious, the dyspeptic, the bloated, the unwieldy, the asthmatic, the lame, and the lazy.

CHAPTER VII.

OPPOSITION TO NEW TRUTHS A CHARACTERISTIC OF THE MEDICAL PROFESSION—ILLUSTRATION SUPPLIED BY THE CASES OF HARVEY, SYDENHAM, PARE, JENNER, HUNTER, SIR CHARLES BELL, AND THE HISTORY OF ANÆSTHETIC AGENTS—THE OPPOSITION TO THE BATH OF THE SAME CHARACTER.

If we reflect on the proverbial inertia of the professional mind, and the indisposition that too generally exists to admit anything which does not accord with preconceived opinion, or commend itself to the accredited dogmas and prejudices of established systems, we cannot be much surprised, notwithstanding the superior enlightenment of our age, that the Hot-Air Bath should have had to encounter, on its revival and establishment, the silent indifference or open hostility of the great bulk of the medical profession—more especially so, indeed, of its so-called heads and leaders. The well-attested therapeutic properties which it presented for investigation were received, shameful to relate, with an ignorant, irrational scepticism by presumptuous guides of medical opinion, who judged by false theories, and obstinately refused to perform the duty of practical inquiry. Yet these very men would eagerly welcome any absurdity if presented in the shape of a novelty in drugging—they would accredit any speculative nonsense in the form of a theory that proposed to whitewash the manifest imperfections in their own empirical art—they would hail with delight any marvelous fiction about drug cures, no matter how self-evident the imposture, provided only that it chimed in with the sanctified routine of thought and practice.

And so has it ever been in all ages of the world; while those who have endeavored to enlighten and improve mankind have suffered persecution, and too frequently been doomed to cruel deaths. "Those who have labored most zealously to improve mankind," says the elder Disraeli, "have been those who have suffered most from ignorance, and the discoverers of new arts and sciences have hardly ever lived to see them adopted by the world." This is true in every department of human knowledge, but in no profession has it been more painfully apparent than in the medical. Its whole history presents a continuous struggle of dominant error to resist the innovations of improving truth. No man whose genius reflected honor on his profession but was compelled to suffer persecution, for no other apparent reason than that his superiority excited the envious feelings of professional contemporaries, who considered they had vested interests in established ignorance and in the mal-practices it sanctioned. This is a truth that stares us in the face on every page of medical history, and it is by thus looking into the proceedings of the past, by observing the conduct of the profession whenever great truths were announced that tended to inform ignorance, increase knowledge, and benefit mankind—by noting the incredulity, the opposition, the virulent abuse they had to encounter, that we can fairly estimate the worth of medical opposition when it is directed in our own day against any such innovating improvement as The Bath, though pronounced a "boon to humanity."

To go no further back than the case of Harvey, who is now held to be "illustrious," in whose honor orations are pronounced, and after whom medical societies are named, we find that his generation did not so recognize his merits. Lecturing unostentatiously on anatomy and surgery to a few students in London, he quietly demonstrated, session after session, to his class for several years, before venturing to publishing it, his grand discovery of the circulation of the blood—a discovery that reveals to us the admirable mechanism of our being, so wonderful in design, so perfect in adaptation, so harmonious in its complexity, and yet so beautiful in its simplicity. But when, in 1628, he did venture to publish his great work, what was the result? Did medical men, anxious to gain knowledge, to acquire any information likely to improve their practice, and enable them more successfully to combat disease, cordially welcome so glorious a discovery? No, indeed; they scouted, and reviled, and repudiated the discovery; and Harvey became the butt of the professional prejudices and presumptuous ignorance of his day.

He was denounced as an impostor, and represented as a quack; ridiculed as "only a dissecter of insects, frogs, and other reptiles," he had to endure whatever opprobrium professional envy and malice could excite against him. We are told "'twas believed by the vulgar he was crack-brained, and all the physicians were against him!"

"The old physicians," says Dr. Pettigrew, in his Biographical Memoirs, "believed that in the practice of medicine there was nothing to be attained beyond what the ancients had already acquired; and they died in the full enjoyment of their ignorance." None, it is said, who had reached forty years could ever be induced to admit the circulation of the blood, or to even study partially Harvey's demonstration! As an excuse for such dogged rejection of new truths, Dr. Johnstone says, "that the physicians of a certain reputation have little to gain, and may lose much'—they pretend, and their credulous patients believe, that their knowledge and practice are perfect; an admission, therefore, that anything could be new and improving to them would be an admission of their own imperfections, which is the last thing empirical practitioners are likely to do as regards the principles or theories of their art. But, adds Doctor Johnstone, "it is shocking that works of great merit have had this misfortune, and Harvey lost patients by his works." When Sir George Ent urged Harvey to print his *De Generatione Animalium*, he replied, "You are not ignorant of the great trouble my lucubrations, formerly published, have raised."

Harvey, however, had a singular triumph—he lived long enough to see his doctrine received and established throughout Europe. "He is the only man I know," observed the philosophic Hobbes, "that, conquering envy, hath established a new doctrine in his lifetime." Yet, as Dr. Pettigrew remarks, "the labor and application of twenty-five years were requisite to ensure the reception of his opinions"—in other words, a generation of old bigots and obstructives had to die out in the full enjoyment of their ignorance before the grand truth of the circulation could obtain general recognition!

The author of "Eminent Medical Men" says, "that even after its truth was generally recognized, Harvey does not appear to have been particularly successful in practice," and the reason assigned is significant—because he disdained those arts of gaining the confidence of the public by which so many succeed." That is, he disdained to pander to public credulities—to cultivate the empirical artifices by which pretentious mediocrity rises and thrives. From the little we can now learn concerning his practice of medicine, it is evident that, as a man of genius, as a skilful anatomist, surgeon, and physiologist, he had a proper contempt for the false theories and gross impostures of mere physic. Aubrey says, "all his profession would allow him to be an excellent anatomist; but I never heard any that admired his therapeutique way. I knew several practitioners in this town (London) that would not give three pence for one of his bills (prescriptions), and that a man could hardly tell by one of his bills what he did aim at. He did not care for chemistry, and was wont to speak against them with undervalue." Evidently he reprobated chemical quackery, and regarded "pharmaceutical filth" as an abomination! Having no faith in physic, as no enlightened physiologist can have, he was too conscientious and honorable to follow medicine as a trade. And even in our own day, no man holding Harvey's opinions, and acting as he did in consistency with them, could attain to eminent repute as a mere physician. In the practice of surgery honorable men can and do attain high distinction by the honorable and conscientious practice of a noble science; but when a mere physician—a drugmonger—gains celebrity, we are at once presented with *prima facie* evidence of the public gullibility, and of his great proficiency in the arts and accomplishments of empiricism.

Sydenham, the most celebrated of English physicians, who has been styled the "English Hippocrates"—a man of great original genius, though he did not always succeed in carrying out or practically illustrating his own admirable principles, yet because he endeavored to elevate medicine from the depravity and corruption in which it was sunk, and establish its practice on the foundation of nature, he was reviled and persecuted by those who considered they held "vested interests" in the gullibility of mankind. He was vituperated as a presumptuous innovator; and he repeatedly complains of the envious opposition with which his labors to introduce a more rational and natural treatment of disease were received by his professional brethren, "who defame their brother," he says, "with this view, that they may gain greater esteem themselves, and build their rise upon the ruins of others; which is a practice utterly unbecoming men of letters, and even the meanest artisans, who have any regard for probity."—Sydenham's Works, Dr. Swan's edit. 3d, 1753, page 117, par. 40.

The calumnies to which Sydenham was exposed from such "envious and malicious men afford, says his editor, Dr. Swan, "a melancholy proof that neither great abilities, unquestionable candor and integrity, nor the most indefatigable endeavors to serve mankind, can secure a person, who leaves the common road, from the unjust censures of the narrow-spirited, disingenuous and prejudiced part of the professors of the same science. Whoever make a new discovery, which tends to overthrow a set of prevailing notions and rules, venerated probably more for their antiquity than justness, and establish a truly rational theory, and more effectual methods of practice, must expect to meet with great opposition from the ignorant, envious, and prepossessed, and be treated as rash innovators, designing and interested persons, however conspicuous they may be for learning, prudence, and extensive humanity."—*Ibid*, p. 117, note.

Similar treatment was experienced by the famous French surgeon, Ambrose Pare, who, toward the end of the sixteenth century, proposed to tie the arteries in amputations with a ligature of silk to prevent hemorrhage, instead of adhering to the barbarous practice, though approved and universal, of searing with red-hot irons. "What!" exclaimed professional prejudice, "allow a man's life to hang upon a thread?" and none of his contemporaries could be induced to abandon the torturing process of the searing iron! So great, indeed, was the reluctance to abandon the cruel cautery for the ligature, that nearly a hundred years afterward a button of vitrol was ordinarily employed in the Paris hospitals for the stoppage of hemorrhage after amputations while Dionis was the first French surgeon who, toward the close of the 17th century, taught and recommended Pare's method.

Jenner, whose discovery has immortalized his name as one of the greatest benefactors of the human race, was assailed with all the envenomed malice his jealous contemporaries could command. "What!" said they, "Vaccinnate!—use such a diabolical invention, and transform the human race into cows and oxen," or into something like the monster described by Ovid—

"Semibovemque Virum, Semivirumque bovem."

To excite the fanaticism of the ignorant multitude against Jenner, it was alleged by London Drug-Doctors that cases had actually occurred, after vaccination, of the whole body having been covered over with a growth resembling cow's hair!—while protuberances that betokened the development of horns and tails had also made their appearance! And as there is no folly which passion and prejudice will not lead men to commit, prints were actually published by the opposing doctors, representing the human visage in the act of transformation, and assuming that of a cow! There was a representation of "Master Joweles, the cow-poxed, ox-cheeked young gentleman;" and of "Miss Mary Ann Lewis, the cow-poxed, cow-manged young lady," who brutified by vaccination, would persist on running on all-fours and lowing in imitation of a cow!

The College of Physicians, who had then a chartered monoply, would not grant Jenner a license to practice medicine in the metropolis! On the contrary, jealous of his fame, he was instigated "to quit London, for there was no knowing what an enraged populace might do!" Even the pulpit thundered at him as a monster of presumption and impiety. A learned divine, Dr. Rowley, declared the small-pox to be "Heaven ordained," and the cow-pox and its use "a daring and profane violation of our holy religion." Jenner and all who favored him were anathematized, and passionate appeals made to excite popular fanaticism—"the projects of these vaccinators seem to bid defiance to Heaven itself, even to the will of God"—was about the most sensible language employed. And thus has it generally been, in all ages, that those who, from their profession and position, should have lighted mankind on the way to knowledge, have on the contrary, labored to perpetuate the deplorable dominion of ignorance and superstition. "Envy," said Jenner in his last days to his friend, Dr. Fosbroke, "is the curse of this country," and among no profession or trade does envy manifest itself in more offensive and dishonorable ways than among high and low who follow physic for its gains. "No science," says the British and Foreign Review, "lulls selfishness to rest—medical science as little as any. Satirists have always delighted to picture the jealousies and personalities of medical practitioners." "Could the London world but know the arts by which certain men have got a name, with what astonishment would it stare," says Dr. Dickson, "to find itself precisely in the position of a deluded savage, when, for the first time, he discovered the utter worthlessness of the red and green glass, for which, year after year, he had been unsuspiciously bartering his wealth. In the dark, pigmies seem giants; Britain only knows her great men when they are dead. On Harvey and Jenner, while they lived, the beams of her warming sun never shone—they all but deferred to acknowledge their merits till she saw them on their deaths, surrounded with that halo of immortality which all the nations of the earth united to bestow on them."—*Fallacies of the Faculty, Preface.*

It is seldom from an ignorant public that great men have received injustice, save, indeed, when popular fanaticism has been inflamed against them by the calumnies of "envious and malicious men" for their own evil purposes. It was from their own profession that Harvey, Pare, and Jenner experienced the grossest injustice, and such has been the case with every distinguished discoverer who preceded or succeeded them.

The celebrated John Hunter, whose life was devoted to anatomical and physiological researches of inestimable value to mankind—who is now eulogized as "the greatest physiologist the world has ever known"—as "one whose labors raised surgery from the servility of a mechanical art to a science of the highest order;" and whose collection of comparative anatomy, which now forms the Museum of the College of Surgeons, London, is represented as "an honor to our age and nation"—as "a monument to his genius, assuidity, and labor, not to be contemplated without surprise and admiration." This truly great man was ridiculed, maligned, and persecuted by the leading practitioners of his day, who were incapable of understanding and appreciating the nature and value of his profound scientific labors. We are informed by Sir Astley Cooper, that a surgeon in London was "hired to write him down!" but he adds, "it was a rat assailing a lion, or a pigmy attacking a giant!"—Life of Sir Astley Cooper, by his nephew, vol. ii., p. 164.

For nearly forty years Sir Charles Bell labored with devoted perseverance to elucidate the mysteries of the nervous system, but at every step his investigations were caviled at, his conclusions disputed, and the merit of his discoveries imputed to others. In the preface to his great work on the nervous system, published in 1830, he says, "the gratification of the inquiry has been very great; the reception by the profession the reverse of what I expected. The early announcement of my occupations failed to draw one encouraging sentence from medical men." He tells us his practice was injured, and that every step in his discovery obliged him to work harder than ever to preserve his reputation as a practitioner. "The prevailing cast of my mind," he wrote to a friend, "was to gain celebrity by science, and this was perhaps the most extravagant fancy of all." M. Pichot, in his life of this eminent man, says, "the vexatious opposition he experienced from the jealousy of his colleagues was such that, when elected to the chair of physiology in the London University, he was soon compelled, for his own peace of mind, to resign the chair. After his removal from London to Edinburgh, and when he had obtained a world-wide reputation, not one of his medical colleagues in the University would call him into consultation, unless forced to do so by the express desire of the patient! He died poor, and Lord Brougham obtained a pension from the Government of £100 a year for his widow.

The history of anæsthetic agents affords abundant evidence to prove that those to whose labors the world is indebted for many substantial blessings seldom have reaped any advantage from them, while we have illustrated the slowness and reluctancy with which the human mind opens to the reception of truth. "There has been a succession of independent spirits," observes Dr. Chapman, "who have refused to acquiesce in the inevitableness of suffering; long before the dawn of organic chemistry, century after century, for two thousand years at least, men have cherished the conviction that, by skillful and patient questioning of Nature, she would be induced to yield up the priceless secret of how pain may be put under the dominion of the human will."

The ancients partially succeeded in causing insensibility to pain, during surgical operations, by various means, but principally by inducing deep sleep from using the root of the mandrake (atropa mandragora). Dr. Snow, in his work on chloroform and other anæsthetics, quotes from Apuleuis, who says: "If any one is to have a limb mutilated, burnt, or sawn, he may drink half an ounce with wine, and, whilst he sleeps, the member may be cut off without any pain or sense." According to Dioscorides, this root was so employed more than eighteen hundred years ago among the Greeks and Romans "to cause insensibility to pain in those who are to be cut or cauterized; for, being thrown into a deep sleep, they do not perceive pain."

To Sir Humphrey Davy belongs the honor of having been the first, in modern times to revive inquiry in this direction, and originating the prolific idea which, says Dr. Chapman, "has at length become one of the most glorious realities of the present century." After ten months' continuous and often hazardous experiments, he published, in 1800, his *Researches on the Respiration of Nitrous Oxide and other Gases*, in which he gave the results of his various experiments; and suggested to the medical profession that, as Nitrous Oxide seemed capable of destroying physical pain, it might be used with advantage during surgical operations. No one, however, thought the subject worth inquiring into. The medical profession paid no attention to his experiments, and Davy's labors went for nothing—were treated as worthless by his generation; and nearly half a century elapsed before any practical effort was made to make such a discovery serviceable to mankind.

In December, 1844, Horace Wells, a surgeon-dentist, of Hartford, Connecticut, in the United States, became acquainted with Nitrous Oxide, from attending a lecture in which the effects of inhaling it were dwelt upon. He immediately conceived the idea of making it available in his practice. He induced the lecturer to go home with him, and to administer the gas to him, while a brother dentist (Doctor Rige) extracted one of his teeth. The lecturer did so, the gas was inhaled, and the tooth drawn without any sense of pain. On recovering from the effects of the gas, Wells was delighted, and exclaimed that a new era in tooth-drawing had dawned. He made several successful experiments on patients; and at last Drs. Warren and Hayward, who were surgeons, reluctantly induced the Medical College of Boston to have a trial made in a serious case of amputating a limb. But, on the appointed day, the operation was deferred; and it was then decided to test the effects of the gas on a person who wanted a tooth drawn. The inhalation was imperfect in some respect; the person alleged he did feel "some pain;" whereupon all the philosophy and logic of the College jumped to the conclusion that the whole thing was an imposture! Wells, overwhelmed with ridicule, returned to Hartford; fell sick through vexation; retired from practice as a dentist; engaged himself in stuffing and exhibiting birds for a livelihood, and in the sale of shower-baths. After many vicissitudes of fortune, his mind gave way; and, becoming more and more unsettled, he finally died by his own hand.

Wells had a pupil named Morton, who settled in Boston, and still cherished the idea of succeeding in effecting painless operations which Wells had failed to establish in practice. In 1818 the celebrated Farady stated that the vapor of æthe—a product known as early as the thirteenth century—produced effects analogous to those which followed the inhalation of nitrous oxide; and Morton, who studied medicine and chemistry under Dr. Jackson, of Boston, learned from him the use of "chloric æther" as a local application. He was thus led to experiment with æther until he became quite confident that he possessed a powerful ænesthetic, when he induced Drs. Warren and Hayward to test its efficacy on patients in the Massachusetts General Hospital. They consented, and, on the 16th of October, 1846, Dr. Warren successfully performed the first painless operation, with this new agent, by removing a tumor in the neck from a patient who remained insensible all the while; and Dr. Hayward, on the following day, was equally successful in extirpating a tumor from the arm of another patient. Morton would not disclose what the agent was that he had so successfully employed; he called it Leothon; but Dr. Bigelow, a surgeon of distinction who was present, suspected, from its odor, that it was sulphuric æther, on testing which he found it produced exactly similar effects, and immediately proclaimed the discovery.

The fame of this mode of abridging the sufferings of humanity spread rapidly in the United States, and as promptly did the instructors and guides of medical opinion take up arms against it—not one of whom alleged that he had taken the trouble of testing the efficacy of the agent, or had any rational idea about the merits of a discovery he so unhesitatingly joined in condemning. Dr. Bigelow relates how the medical journals combined to write it down in the face of the undoubted success that had attended its use. A leading medical journal in New York announced that:

"The last special wonder has already arrived at the natural term of its existence. It has descended to the bottom of that abyss which has already engulphed so many of its predecessor-novelties, but which continues also to gape for mor until a humbug yet more prime shall be thrown into it!"

A medical periodical of repute in Philadelphia declared:

"We should not consider it entitled to the least notice, but that we perceive, by a Boston journal, that prominent members of the profession have been caught in its meshes"—distinguished surgeons who had tested its merits, and gladly hailed it as a "boon to humanity." Yet this high physic authority "was fully persuaded that the surgeons of Philadelphia would not be seduced from the high professional path of duty into the quagmire of quackery by this will-o'-the-wisp."

"That the leading surgeons of Boston could be captivated by such an invention has excited our amazement," wrote a New Orleans representative of the Drug School; and by such unreasoning invective, it was sought to discountenance and put down this invaluable discovery. The same incredulity and opposition attended its recognition in Europe as in America. In November, 1846, Dr. Bigelow states that information of the discovery was sent to London and Paris, but he says the distinguished surgeons of Paris received the announcement with indifference. Even the great Velpeau "politely declined" to test its worth! Dr. Boot, of London, who was personally acquainted with Dr. Bigelow, was induced by his representations to make a trial of "the new anodyne process," which he did, on the 19th of December, with perfect success. Dr. Boot then sent Dr. Bigelow's letter to the celebrated Liston, entreating him to test its merits; and, on the 21st of December, this great surgeon, at University College Hospital, was most successful in its use. In a letter written next day to Professor Miller of Edinburgh, prefaced by exclamations of his intense delight—"hurrah! rejoice!"—he gives an account of the operations he performed—he "amputated a thigh, and removed by evulsion (forcibly plucking out) both sides of the great toe-nail, without the patient being aware of what was doing, so far as regards pain, The amputation-man heard, he says, what we said, and was conscious; but felt neither the pain of the incisions nor that of tying the vessels."

What the interests of humanity, and a love of scientific progress could not induce the medical profession, with its pride of opinion, its prejudices and its bigotries, to do, the celebrity of Liston, the unquestionable authority of his name, at once accomplished, viz., secured for the "new anodyne" a fair investigation of its merits. Experiments were forthwith instituted in nearly every hospital, and, though in some instances with unsatisfactory results, owing to ignorance in properly administering it, yet the general success was undeniable. The evidence thus accumulated both in America and this country, shamed, as it were, the two great French surgeons, Velpeau and Roux, into testing its merits; and in January, 1847, they declared, in presence of the two Academies of Paris, that their investigations had been eminently successful, and that the discovery "was a glorious conquest for humanity."

It must not be supposed, however, that the repugnance to what is new and improving, and for the advantage of mankind, which is so marked a characteristic of the medical profession, was in any way abated as regards the employment of anæsthetic agents. On the contrary, an absurd theory was forthwith invented—when the power of these agents could be no longer denied—by which opposition to their use was sought to be justified on the ground that the feeling of pain was essential to the healing process! Men of high repute were absolutely found to contend that pain is salutary! that when a patient is undergoing some excruciating operation, the agony he is compelled to endure is exceedingly good for him! and that, consequently, the annihilation of a sense of pain is likely to prove highly hazardous to the patient's life! This theory, so repugnant to sound physiology, found a supporter in the celebrated Majendie, who "doubted if there was any advantage in suppressing pain"—who averred "it was a trivial matter to suffer," and that "a discovery the object of which was to prevent pain, was of a slight interest!" Such a thoughtless expression of opinion was totally unworthy of so eminent a physiologist; yet it found favor with a lot of practitioners who had no desire to depart from established routine by learning anything new. So difficult, indeed, is it to uproot inveterate prejudices in the medical mind, that even so late as the Russian War, the Director-General of the Army Medical Department would not sanction the use of chloroform in operations—the sense of excruciating pain he barbarously considered was a salutary adjuvant to the healing process!

We have brought these few illustrations of medical obstructiveness before the reader, in order that a just estimate may be formed of the value of medical opinion when pronounced in opposition to any discovery that promises to be of advantage to mankind. Men who have the intellectual capacity and true philosophical spirit to investigate and think for themselves, when they proclaim truths distasteful to professional ignorance and prejudice, must be prepared to encounter the rancorous vituperation of "envious and malicious men," who, caring nothing about truth, never dream of forming an independent opinion, and detest those who do. "Persons who object to a proposition merely because it is new," observes Sir Astley Cooper, "or who endeavor to detract from the merit of the man who first gives efficacy to a new idea by demonstrating its usefulness and applicability, are foolish, unmanly, envious, and illiberal objectors; they are unworthy of the designation either of professional men or gentlemen."

No one even superficially acquainted with medical history could, indeed, be guilty of the folly of judging the merits of any newly-discovered remedial agency, that was repugnant to medical dogma, by the opinions which even the most eminent of the profession might express in detraction of it. The Professors, the standard writers, and those who inspire the periodical literature of the profession, are generally the most bigoted, the most intolerant, and the most virulent against new and improving truths. "Two sorts of learned men there are," observes Bishop Berkeley: "one who candidly seek truth by rational means; these are never averse to have their principles looked into and examined by the test of reason. Another sort there is, who learn by rote a set of principles, and a way of thinking which happens to be in vogue; these betray themselves by their anger and surprise, whenever their principles are freely canvassed." And equally so, when anything new is proposed that is not dreamt of in their philosophy.

Just as there are no devices of charlatanism so gross, no errors so pernicious, that medical men with very learned pretensions and high repute have not been found ready with plausible arguments to uphold and defend, when they were in accord with their preconceived opinions, so has it ever been with Truth—no great truth, no useful discovery ever yet obtruded on established dogma to disturb the satisfied repose of monotonous routine that was not unhesitatingly denounced and, if possible, persecuted as dangerous to society, while those whose object is to obstruct human progress never yet were at a loss for colorable reasons to justify their wickedness. But "the asserter of truth may be crushed, and we may breathe a sigh over the martyr as he passes from the field of his labors—ignorance and prejudice may for a time reign triumphant, and the abettors of sloth and selfishness be considered the great, the good, and the wise—but Time rolls on, and Reason will assert her dominion."

The medical profession, of all others, ought to be inspired with lofty aims, and prompted by noble impulses; but its vision is too often bounded by a narrow selfishness, and its motives are too frequently groveling. The enlightened and truth-seeking, the men who labored like Harvey, or Hunter, or Bell, receive no appreciative sympathy from those who are interested in retaining things as they exist, and have no desire to cast aside erroneous opinions even to make room for truth. Such a spirit is not only hostile to the progresss of science, but it tends to the positive injury of mankind, by increasing personal suffering from preventible disease. Yet in such a spirit was the establishment of the Bath received by the great bulk of the medical profession—by men, too, who never pretended they had studied or tested what they so rashly presumed to condemn, no more than did their worthy predecessors pretend to base their opposition to the discoveries we have noticed on any knowledge of their merits, but blindly opposed, just as their pride, or passion, their ignorance, interests, or prejudice prompted them. What that opposition was worth, experience has now proved; and as we marvel at the intolerable perverseness that dictated it, so will the next generation wonder at the stolid prejudice and folly of the present in its opposition to such an undoubted therapeutic agency as the Bath.

For the most part, however, the objections that have been urged against the Bath are of a very frivolous character, and not over-worthy of serious consideration; but a notice of them may be made useful, and prove satisfactory to those who desire full information on the subject, and who may still be inclined to think that an objection must have some reason in it because it is advanced by a member of the medical profession.

The Turkish Bath--Its Origin and History.

LECTURE BY COL. T. T. GANTT.

A DESCRIPTION OF THE BATH—ARGUMENT FOR ITS INTRODUCTION IN ST. LOUIS—ITS IMPORTANCE AS A CURATIVE AND PREVENTIVE AGENT, ETC.

A large and select audience gathered in the hall of the Polytechnic building on Saturday evening, to listen to the lecture of Colonel T. T. Gantt, on the Turkish Bath. The following is the lecture in full, which is of a most interesting character:

I propose to address you respecting what is called the Turkish Bath. There is something misleading in the name. One is apt to suppose that it signifies an institution having its origin with the Turks. Now, the Turks were, before the taking of Constantinople, in the middle of the fifteenth century, destitute of the bath, and were a filthy people in their native land. When they took Constantinople and overthrew the Roman Empire, of which that city was the metropolis, they found the bath there. The seat of the Empire of the Cæsars was transferred to that city from Rome by Constantine in the early part of the fourth century. With the Court, he transported to the shores of the Bosphorus the habits and the institutions of the Romans. But for long years before the time of Constantine the thermæ of Rome had been one of her most cherished possessions. We are told by Tacitus that the nobles of the Imperial City, both of his day and of the previous ages, considered the daily use of the bath (as we translate the word thermæ) to be not only the best preserver of health, but the surest barrier to the inroads of old age. Was it, then, a habit, an institution, indigenous at Rome? Its name, thermæ, which is the Greek word for heat, tells us that the Romans received the bath, as they did so much else that was useful, ennobling and elevating, from that most wonderful of all people, with whose history we possess even a tolerable acquaintance. But whence did the Greeks derive it? Was it a discovery of their own, or did they borrow it from an earlier civilization? We have the best reason for believing that they derived it from the Persians, and they again from that earlier and almost mythical race, the Aryans, to which modern researches point as the source of much that is most striking and precious in human history. But it would be a more than useless excursion if I should pursue this topic further. I only meant to say that this which we call the Turkish Bath is so called because, after being taken to Constantinople by the Romans, who under Constantine fixed the seat of Government at that city, it was found there by the Turks, who, about the middle of the fifteenth century, took possession of that place by conquest, and there established their own empire. The Turks, who were, perhaps, incapable of originating such an institution, were sensible enough to adopt it; and by this adoption it has continued for more than four centuries. The provinces of Turkey, including that memorable land known to us as Ancient Greece, retained this health-giving agent until 1829. Then, the independence of Greece being accomplished, mainly by the assistance of the English and Russian fleets at Navarino, emancipated Greece threw off everything, or nearly everything, that could remind its people of the hated Ottoman yoke. The rejection was too sweeping to be quite wise. No doubt some bad things were rejected, some evil habits which were common to Turks and Greeks alike; but also some good things came to be thrown aside merely because they were Turkish, and some bad or inferior habits were adopted merely because they were characteristic of their Frankish liberators. Of these last was the bath. The bath as then used at Constantinople, throughout the provinces of Turkey and in Greece itself, was the most perfect thing of the kind ever known to man. The bath as used by the Europeans, and as generally of that period known to us in America, was altogether inferior, both in its pleasurable and health-giving influences. Yet such is the potency of prejudice and fashion, that under their combined action the thermæ of their fathers disappeared in Greece, and the baths of the Franks usurped their place. Turkey was thus left the last asylum of this wonderful and wonder-working agency; and it is said that since the assistance rendered to the Turks by the French and English during the Crimean war, so fashionable have Frankish habits become, merely as such, in the capital of their empire, that the true Turkish Bath is there replaced by the European, and the peculiar conditions and practices which give all its value to the ancient thermæ are now only to be found in the provinces of Turkey.

But here some compensation begins. For years and years Christians visiting Turkey from Western Europe enjoyed the heat, the manipulation, and the perfect cleansing of the thermæ without ever dreaming (apparently) that the same agencies might be made as active and beneficial on the banks of the Thames, the Severn and the Seine as on the shores of the Bosphorus. About one hundred and fifty years ago, Lady Mary Wortley Montague, in her admirable letters, described the manifold blessings and luxuries of the bath of the Turks. But it was reserved for Mr. Urquhart, an English gentleman, who spent many years in Asia Minor and Egypt, to bring home with him a determination to extend to his own countrymen the luxuries and the blessings of this bath. For years he has devoted himself to the promotion of this enterprise. Like all benefactors of the human race, he has reaped a plentiful harvest of ridicule, of opposition, and even of obloquy. He has, however, triumphed over all these, and now enjoys the proud satisfaction of knowing that the most distinguished of the medical men of Great Britain and Ireland recognize in the Turkish Bath one of the most powerful weapons wherewith to combat disease, whether in its insidious approach or after it has effected what, but for its operation, would be a fatal lodgment in the vitals of the human frame; that for many of the ailments of modern times it affords the only relief which is known to science; and that for strengthening the several organs of sense, and invigorating and elevating the general tone of the system, it is absolutely without a rival. When to this we add that in its administration it is a source of the most pleasurable sensations; that as a cleanser of the body it stands pre-eminent, and that the mind, as well as the body, partakes of its refreshments, we must be indifferent to what would naturally arrest the attention of the most heedless if we did not wish to become personally acquainted with an agent of such beneficent influences. I am here this evening to detail to you in words the operation of the Turkish Bath. I will, step by step, conduct you through the several processes; and I will then ask you to consider whether we are wise in permitting all these advantages to escape us for the want of a little exertion to make them our own.

The building devoted to the Turkish Bath necessarily contains—1st. A reception room, which may be small, for any one who comes may at once pass into the dressing-room; and it is not proposed to furnish a place for lounging. 2d. A number of small dressing-rooms or closets. 3d. A room commonly called the tepidarium, the temperature of which is from 120 to 130 deg. Farenheit. 4th. A room called the calidarium, the temperature of which is from 145 to 180 deg. of Fahrenheit. 5th. A smaller room, furnished with marble slabs, about seven feet long by three feet wide, and also with pipes supplying cold and warm water. 6th. A tank or vat, which should be from four and a half to five feet deep, fifteen feet long, and ten feet wide, but which may be much less. This tank should be constantly emptied and replenished by a flow of clear water. Until our hydrant water is furnished to us free from its sediment, this part of the establishment must remain in abeyance in St. Louis. 7th. The circuit is completed when we pass into the cooling-room.

On reaching the bath-house, the bather proceeds to one of the dressing closets, and completely disrobes. He is then furnished with a piece of chintz, a yard wide and several yards in length, which is tied round the loins; and it is, in my opinion, well to add to this costume slippers made of straw, or wooden pattens. The reason for this you will perceive presently. Attired thus, the bather proceeds under the direction of an attendant to the tepidarium, which is heated, as before stated, to 120 or 130 degrees Fahrenheit. The floor of this apartment is laid in tiles. In some baths the heat is applied from below; in others it is radiated from pipes fixed at the sides or at one side of the room. In the first case it is absolutely necessary that you should have the feet covered by some non-conducting material, even for the short period that you occupy in walking to your chair. The heat will otherwise be quite intolerable, even when the heat is radiated from pipes of fire-clay. The tiled floor is, in the calidarium at least, so hot as to be very trying to the sole. You are seated in a cane-seat chair (on which a clean sheet is spread), and furnished also with support for the feet. The attendant places a napkin dripping with water on your head, and you are left for a space to your reflections. Your sensations are very agreeable. The room is not dark, but obscurely lighted. You see, perhaps, as many as a dozen others similarly situated, but are unable to distinguish anything beyond the fact that you are not alone. If you desire anything from the attendant you clap your hands. Silence, decorous silence, is the rule of the bath. It is one of the good things that we have borrowed from that unloquacious people, the Turks.

At the end of fifteen minutes (more or less, according to circumstances) the attendant, finding that perspiration has freely commenced, conducts you into the next room, the calidarium, heated, as before stated, from 145 to 180 degrees. Those parts of the room nearest the source of heat will naturally be the warmest, and the different temperatures I have named may be found at the same time in different parts of the same room. Here you are seated as before, and the perspiration which began in the tepidarium goes on vigorously in the calidarium. Indeed, such streams as flow from you are, if you have never been in a calidarium before, a contradiction to all former experience. This will not surprise those who remember that the hottest bath of water can not be carried beyond 120 degrees at furthest, and that few can endure it beyond 115 degrees. Not only without inconvenience, but with positive enjoyment, you now revel in a temperature at least 40 degrees beyond that point. The difference is, that the medium now surrounding your body is dry electric air, not hot water. Your breathing, if you labor under any dyspnœa or difficulty of respiration, is greatly relieved. If you are troubled with pulmonary disease, you at once perceive the balmy influence of the dry, warm atmosphere you inhale. Your breathing is soft, and not at all hurried. The action of the heart is slightly accelerated, and yet, though accelerated, there is nothing irregular in it, and one of the cases in which the Turkish Bath is most benign is that of heart disease.

The grand feature, however, is the thorough, complete diaphoresis (that is the medical term) which is going on. You almost seem to melt away. Very probably you feel thirsty. If so, you signify your want, and the attendant hands you a glass of cold water. You swallow it, and presently it pours out of your body through your skin, every pore of which is open and streaming. I ask your attention to what is now going on and has been going on since you entered the tiled chambers. The perspiration of any of us on first going in would be highly acrid, carbonized and saline. As time flows on, and the perspiration continues to pour, particularly if the bather drinks freely of water, the perspiration loses its saline character. It is insipid as spring water. Now, whence does this perspiration come to the surface of the body? Every one knows that it is given off by the blood. It follows, then, that the blood, as the process advances, is being freed from the acrid ingredients which characterize our first contributions. Urea, carbonic acid and salty matters, mixed with a fatty substance, were in the perspiration of the first half-hour. During the succeeding thirty minutes these deleterious elements will have almost entirely disappeared. It is the business, the function of healthy perspiration, induced by exercise, to eliminate these noxious products from our blood. If we strictly observed the rules of temperance; used only proper clothing, and took plenty of active exercise in the open air, keeping the skin in a normal condition, the necessity for the thermæ would be lessened. None of us do this, however, or not one in a million. The system then becomes loaded with the products of the vital action of our organs. The insensible perspiration which constantly brings to the surface a large amount of these effete substances, instead of passing into air by a volatilization, is checked and collected as a varnish on the skin, dangerously and sometimes fatally hindering its proper and beneficial activity. I have been told that there are some persons who consider this greasy, sour accumulation of carbonized matter mixed with urea and dead skin, to be a shield against atmospheric changes which it would be imprudent to remove by washing! All such remarkable people may make themselves easy, and wash without fear, for two reasons: First, The more of this filthy covering they remove, the less poison will be in danger of being re-absorbed into the system, and the freer will the skin be made to do its appropriate work as an emunctory. But secondly, This abominable plaster, or varnish,* is only very partially removable by the best of scrubbing with brushes, soap and hot water.

* Not only is this varnish capable of making us disordered; it only needs to be quite impervious to air and moisture to be speedily fatal. At a grand pageant many years ago in Europe, it was thought expedient and becoming to symbolize several things—among others the Golden Age. The representative of this good old time was a little boy, furnished with a crown and wings and some drapery. In order to make him properly representative, his body was covered with gold-leaf. The ceremony lasted several hours. Before they ended the Golden Age had died of asphyxia or suffocation.

When these have done their best, the person remains far from clean, and of this truth the next step in the thermæ gives demonstrations; for having passed a sufficient time, varying with inclination, leisure, health, etc., from forty minutes to one hour or more in the calidarium, you are conducted into the fifth apartment (generally separated from the calidarium by a curtain only), and placed on a slab of marble to be shampooed. This operation is performed by an attendant, who goes over the course of every muscle of the body with considerable pressure of the hand. Now it is that a man fresh from daily bathing in the surf of the ocean, though he also has rubbed his body strongly with the sand of the beach at each tide, or confident of being clean beyond challenge from the daily use of the bath at home, with all the aid of brushes, soap and sponges, awakes to the mortifying consciousness that up to this day he has never been otherwise than very dirty. The dead skin which collects under the hand of the shampooer and rolls down by your side like something between vermicelli and maccaroni, is absolutely what must be seen in order to be believed. How much of this comes off at the first operation, I do not precisely know. But it is laid down by those who should be acquainted with their subject, that as much of this dead skin forms on the surface of the body each week, and is there compacted into a paste with the unfragrant products of perspiration already mentioned, as will, when removed by the shampooer and dried, form a ball about as large as the fist of the individual.

The effect of the shampooing following the great heat, is to render the whole body supple, to remove stiffness of all kinds, and to banish pains as if by magic. Rheumatism, gout and neuralgia are said to be completely relieved by it. I can answer for the truth of the saying as far as rheumatism is concerned. If you had a catarrh, you left it in the calidarium. You have more than attained, without fatigue, all the effects of the most prolonged exercise. You have effected such a "purgation of the blood through the skin" as could be brought about by no amount of exercise that the human frame could endure. You have enjoyed, without disturbance of a single organ, such a diaphoresis as no sudorific medicine could accomplish without great risk to the constitution. In a word, this application of heat, and the combination with it of the manipulations I have described, accomplishes in a most marvelous manner all that the most regular and persistent exercise can effect, and all this without the least fatigue or strain.

But the business is not ended. When all the dead skin is removed, and when every muscle has had enough of that hand-rubbing, which we have so long known to be useful for our horses, but which, until very lately, no one seemed to think of extending to our own bodies, the whole surface is rubbed with soap on a stiff sponge or a hair glove. Now comes the time for the first application of water in this marvelous bath. Water here is only used for rinsing and cooling. The cleansing is effected by the sweating and rubbing. You are laid on your back on the slab, and the attendant dashes a bucket of warm water over you from the head towards the feet. He turns you over and repeats the operation on the other side. He then raises you up, and if there is a tank, and you desire to plunge, he leads you to it. If there be no tank, or if you do not care for a plunge, he places you under the shower-bath, and turns on first water which I always considered unpleasantly warm. He makes it colder to your taste, and fixing the run of water from above, he at the same time turns on you a douche from a small nozzle. You turn round, so as to expose each part of your body to this stream, and when sufficiently cooled, you are dried with towels, enveloped in a dry sheet, and conducted to the cooling room, there to sit until you are ready to dress yourself. You may fancy yourself ready to do this the moment you come within reach of your clothes. You are mistaken. If you dress at once perspiration will re-commence, which will be inconvenient certainly, and possibly injurious. The pores of the skin have been braced and closed by the dash of cold water. But such is the vigor imparted to every vital function by the process through which you have passed that it requires some interval before you can support the confinement and incumbrance of your customary integuments. After sitting for a space, varying from fifteen minutes to half an hour, undressed in an atmosphere which, when you first came to the establishment, was pleasant to you with your usual clothes, you begin to feel that they would be comfortable once more. At the first intimation of this truth you don them; and may then proceed to such duty or pleasure as belongs to the hour.

The question now arises, what effect has been produced on the bather? He is refreshed, strengthened, fortified and exhilarated. He is purified from all uncleanness. He is able to endure with impunity an exposure to atmospheric vicissitudes, which a few hours previous would have caused congestion and inflammation—catarrh or a pleurisy. He is relieved from all obstructions. His vision is clearer, his step more free and elastic, his mind more cheerful. To all that dreary train of symptoms commonly described by the term bilious he bade adieu in the calidarium and the shampooing room. He has not been subjected to the disturbing influences of a cathartic, a diuretic, or a sudorific operation to purify the blood through the stomach and intestinal canal, nor has he been enabled to perspire only by exhaustion and painful exertion; but while he has been almost wholly passive, a "purgation through the skin," as it was termed by Hippocrates, the most revered name in medicine, has eliminated every effete and noxious particle from his circulation.

I can not forbear calling your attention to this marvelous result. We saw the process going on in the calidarium. We saw the acrid, carbonized perspiration giving place to a discharge as limpid and almost as pure as spring water. It is told by Dr. Rush, one of the foremost names in American physic, that when in 1820 the yellow fever raged in Philadelphia, there seemed to be some local poison which infected slowly all who remained in the city, and all who came to it from abroad. What may be termed "bilious symptoms" manifested themselves in these persons some time before the actual breaking out of the disease. If in this condition the blood of any of these persons was drawn and examined, it presented, though in a less decisive degree, the marks of disorder which were exhibited by the blood of those actually laboring under an attack of the yellow fever (for Dr. Rush bled his yellow fever patients. Modern doctors know better). At this juncture, which may be called the premonitory symptoms of the disease, it was of course easier to treat and baffle it than when it had advanced to full development. But does it not seem inevitable that if a person in this condition could have been enabled to cast off the diseased elements of his blood—if the noxious particles could be eliminated by such a "purgation through the skin" as has been described, that, without impairing the strength, the healthful condition of the system would be restored as by a charm? Remember that the constant effort of nature, by all her emunctories, is to throw off that which may be called excrementitious matter. Every thing that we take to sustain life, the air we breathe, the food we eat, and the drink we swallow, serves its purpose in the vital economy and becomes decomposed in the laboratory of the assimilating organs. So much as is capable of entering into our organism is retained. The rest must be expelled, or it will poison us. There is a constant process of expelling poison going on during the healthy action of our organs, and the wisdom of the Creator has so wonderfully and fearfully formed us, that what are called diseases are after all only the efforts of nature to cast out poison of one kind or another. Thus, we are poisoned by miasma; and nature, seeking to eliminate this poison, raises the temperature of the body to what is called fever heat, and the result is either a copious perspiration, which relieves at least temporarily, or else congestion and death. We take into our stomachs a mineral poison, and nature seeks at once to reject it by a convulsive effort. We swallow, for example, what is not a poison, but in many cases a valuable medicine, the hydriodate of potash, and in an incredibly short space of time it has been taken into the circulation and is discharged by the skin. The air we breathe becomes changed in the process of respiration. It entered the lungs composed of oxygen and nitrogen. It leaves the lungs composed almost wholly of nitrogen and carbonic acid gas. When breathed in, it was capable of sustaining life; when breathed out, it is literally a poison. So it is with our food and our drink. And so it is with the matter that is thrown off by the pores of the skin. It is simply poisonous, or at lowest, noxious matter. When we exercise freely and perspire copiously, this elimination is much more perfect than when we lead sedentary lives. And this is, in a few words, the main utility of exercise in the open air. But it is not enough that this peccant matter should be brought to the surface of the body. If left there, it will clog the pores of the skin by its unctuous, fatty substance, and, combining with the dead skin (which may be said to be in a constant state of dying at the surface), will form a varnish, which will almost entirely put a stop to the insensible perspiration, besides being re-absorbed into the system to ruin our health. I will say nothing of the filthiness of such a habit of body, but who can fail to see how conducive it is to disease, especially to headache, pulmonary complaints, and affections of the liver and kidneys?

I have been asked if the excessive perspiration induced by the bath is not weakening. Men are so much in the habit of associating perspiration either with severe toil, or with the sensation of nausea, both of which are amongst the most exhausting things in nature, that they expect all perspiration to be debilitating, and, as it were, prepare themselves to feel weak when they sweat inordinately. They persuade themselves that it *must* be so, and their imagination comes to their aid and makes them believe that it really *is* so. We have all heard of the celebrated anecdote of the criminal condemned to be bled to death, whose eyes were bandaged, while a scratch was made in the neighborhood of a vein. Warm water was then poured over the part, and he heard the attendants (medical men) commenting on his failing pulse, and the emptying of his veins. They went on to describe, in his hearing, the symptoms of approaching death by loss of blood, and the victim, who had not lost an ounce of it, actually perished by the force of imagination. So at least runs the story. Something of this kind seems to me to have been the experience of a few nervous men who have taken the bath. For my own part, when in New York last June, in eight days I took six Turkish baths. I usually took them in the morning before breakfast, but in some cases toward three in the afternoon, and always before a meal; sometimes before breakfast, and sometimes before dinner. I was invigorated and strengthened by each of them. I was delighted with the sensations they produced; I was greatly benefited by their action; and if I were to be critical at all, I should say that I would prefer the heat of the calidarium to be greater than 165 degrees (which was the hottest I could procure), the shampooing to be more vigorous than it was, and the douche to be colder. It is the experience I have had in my own person of the benefits resulting from this process that makes me desire so ardently to bring them within my daily reach, and to do this not merely for myself, but for all my fellow-citizens.

This brings me to the last topic which I propose to discuss. I have not said a tithe of what I might say, if your patience and my power of utterance would allow, on this (to me) most interesting subject. But I must pass on to the question: "How are we to obtain such a bath here?" And to this question I have but one answer. It can only be done, humanly speaking, by co-operation; and, in order to secure co-operation, the minds of those whose assistance is needed must be convinced of the utility of the scheme. The great difficulty is to awake the attention of our citizens to it. This is no easy matter. There is such a seductive ease in ridiculing the enterprise, compared with the attention necessary to understand its merits, that we have, perhaps, no right to be surprised that a majority take the course first indicated. He who ridicules, again, and expresses doubts of the success of effort, appears to his own eyes to be wiser or shrewder than he who is believing and working. With all these obstacles we must struggle, and we can only succeed when we have conquered and overthrown them.

The first step is to induce a general belief in the utility and desirableness of the Turkish Bath. If this can be effected, it really seems to me that all the rest follows. If it were felt as a conviction that this bath would not only avert or cure some of the most disabling, tormenting and fatal diseases of our climate, but that it was a source of the most lively enjoyment to the healthy, there would be a general determination to possess such a talisman; and no terms that could be named would be likely to be regarded as exorbitant when weighed against such advantages. If we were the subjects of an emperor, being ourselves slaves, we would naturally expect that if we had the baths at all, we must owe it to the bounty of our sovereign. The masters of Rome felt and fulfilled the duty of making this provision for the populace. Scarcely a single emperor failed to enrich the Imperial City with some structure dedicated to this beneficent end. One of them — Diocletian — constructed one of these thermæ capable of accommodating at once 3,000 persons.

While Rome was yet a republic, thermæ less magnificent and spacious than those of the empire, but still far surpassing in both respects anything in the United States, or perhaps in modern Europe, were constructed by the wealthy nobles and made free to every citizen. This was also done in Athens. It was one of the old fashioned modes in which men of immense wealth sought to employ it. The practice has gone sadly out of vogue, and besides there is scarcely any example now of that accumulation of wealth in the hands of a few, which made such magnificent liberality possible in the ancient world.

If we want the Turkish Bath we must build it ourselves. It is hardly possible for any private individual to possess such a bath on his own premises unless wealth renders expense of no consequence. And for many reasons it is expedient that the thermæ of St. Louis, if we ever have such a thing, should be accessible to all, and be divided into two great divisions, one entirely for men, the other entirely for women. Of one thing I am confident, if we succeed in having one of these establishments, there are those living who will see enough of them to accomodate 5,000 bathers a day. It is sickening to think of the difference in our health, comfort and substantial wealth, if the money which has been wasted in building, pulling down, remodeling and rebuilding the courthouse, for example (not to speak of other wastings of our money), had been expended in giving to the city an adequate number of thermæ. But such reflections will not bring back the money, nor make the plunderers of the public treasury yield up their spoil. We must ourselves, *now*, contribute enough for at least a beginning, or a beginning we shall never have.

The least sum sufficient to make the first payment on a suitable lot of ground, and to erect on it buildings with appropriate apparatus for a bath sufficient to accommodate both male and female bathers at the same time, is about forty thousand dollars. For this sum we may have a bath of which we may be reasonably proud, because it will adequately meet our immediate wants. There will be nothing *grand* about it. But yet it will compare favorably, perhaps, with any similar establishment in the country—so poorly are we, citizens of the modern republic, in this enlightened nineteenth century, furnished with what many nations of the old world, whom we are not ashamed to speak of as barbarians, possessed in profusion and magnificence, free to every citizen and considered as an indispensable condition of health and comfort. An effort has been for some time past making in St. Louis to raise this sum. Some few have subscribed $500 apiece toward it. Perhaps $15,000 has been thus raised, provisionally, for it is a condition of the scheme that no subscription shall be binding until the whole sum shall be subscribed, and in the last month not a thousand dollars has been added. It is quite plain, then, that the enterprise will fail unless some assistance, hitherto withheld, shall be given. And it is with the purpose of exhorting you not to permit it thus to fail, that I am before you this evening. Let no one imagine that the Bath will be built anyhow, and that he may fold his arms while the work is done by others. The effort of raising the requisite funds has been going on since the middle of September last, and, by all appearances, we have realized a most trying disappointment. Unless something more encouraging than our recent experience will reward our exertions, they will hardly be persevered in, and, the whole business will fall through. I have been, for one, active in canvassing for aid to the enterprise. I have met with many disheartening replies. I have been asked, when I have spoken of the inestimable benefits anticipated from it to the health and improved habits of our people, whether it will pay a large profit to the stockholders or subscribers. Health, cleanliness, comfort, alacrity of mind and body, are too little, it seems; will dividends follow also?

Of course, the idea of the questioner is, that non-subscribers will get as much of the benefits to health as subscribers, and that the subscriber who puts up his money ought to get something additional for it; and in this there is a show of reason, if we leave public spirit entirely out of view, and deny that such a thing exists. But, in the first place, I may say to some one who is able to spare a subscription, without reference to any remitting dividend, but who hesitates on this ground, that if the rest of the world, in his situation, think and act on this principle, we shall have no bath at all. In the second place, it will be an excellent investment for any one, if by subscribing, for example, $500, $250, or $100, he can promote the establishment of a bath at which he can, whenever he pleases, cleanse his body, purify his blood, and renovate his health at the same time that he experiences the most pleasant sensations, even though otherwise his subscription is sunk, or, at any rate, pays no money dividends. And it certainly seems to me to be a pitiful thing that any one who believes the bath to be capable of these beneficent results, and who can afford to pay for the necessaries of life (I should be glad to be informed what deserves to be called such, if health and cleanliness do not), can stop to inquire whether there is money in it. At what time in the millennium will men understand that a penny's worth of health and comfort is well worth a penny? But in the third and last place I will say, and I am almost sorry to say it, that I believe there is money in it. In other places the Turkish Bath has paid handsome dividends. In no place is it more needed than in St. Louis. There is a gentleman here familiar with the management, who in view of what seems to him an immense field for its operation, stands ready to pay the association that will build such a bath as I have mentioned, fifteen per cent. annually upon the outlay. He makes this offer for the first year, leaving it to the experience of that year to determine what it will be prudent or necessary to offer for those that follow. Therefore, although I deprecate this test of the propriety of engaging in the enterprise, I think I may say that it will most certainly not pay merely in health and comfort, but in money, too. * * * I have not said as yet a word to the ladies, yet none are so deeply interested as they in the success of this enterprise. Nothing is more commonly said than that the climate of St. Louis is very unhealthy for ladies; and when it is replied that this is owing to the fact that they lead such sedentary lives, the rejoinder is, that the climate is so debilitating that they can not exercise. One objection to this denunciation of our climate is that men and women both belong to the human species, and that it is odd that a climate not unhealthy or debilitating for men should be so for women. Another objection, is that there lived here in the last century and the present some ladies who retained their health and activity to a very advanced age; while the corresponding fact in relation to men is too notorious to need repetition. But it stands admitted on all hands that the ladies of St. Louis would enjoy good health if they took exercise. Well, the results of exercise are attainable by means of the thermæ, without fatigue; and this is only another mode of saying that the Turkish Bath will give health to our female population. It will restore that clearness of complexion, that bloom of color which, as is an infallible index of health, is one of the highest beauties of women. In fact, for the fair sex, health means beauty. English women owe to their resolute, active habits, their determination to walk abroad and breathe the fresh air in all weather, that permanence of their good looks, the want of which is so deplorable with their American sisters. But by the adoption of this substitute I feel satisfied from every rule of analogy that they will be able to retain, late in middle age, that brilliancy of personal charms, the possession of which by our women during the period of extreme youth is universally admitted. Now, I beseech the ladies to remember that full significance of this change. Their complexions fade now, because their health fails. If the health is fortified and preserved, they will retain their beauty. But I have too much respect for them to speak as if the mere gratification of their vanity would be a sufficient motive to make them strongly desire the establishment of the Turkish Bath. It is because with improved health will come an enlarged capacity for enjoyment and usefulness, and a freedom from that disabling weakness, languor and helplessness that, with too many, make life a burden. It is for these reasons that I desire this bath for them, and would have them desire it for themselves. Here permit me to remark that with my good will no such establishment shall be built in St. Louis unless it provides equally for both sexes. In all discussions that have taken place, not only I, but all who have testified their willingness to co-operate with me in this matter, have repudiated the idea of but one suite of rooms with their appliances, open to men at certain hours, and to women at certain other hours of the day. Our scheme is different. We propose to have a building divided in the centre. On each side of this there will be a complete and perfectly equal bathing apparatus. The attendants in one of these will be men, and into this division no woman is to enter at any time. I consider it of inestimable advantage to have this therapeutic agent under the charge and direction of the medical profession. It will insure us against that quackery which is the deadly foe of all improvement in the means of cure. The agent here employed—heat, dry and electric, combined with manipulation—was known and approved more than four centuries before our era by one of the greatest men the world has seen; the man whom preceding ages revere as the father of the art of healing.

I have, therefore, omitted all citation of the opinion of medical men of eminence in Great Britain touching the efficacy of the Turkish Bath in the prevention and cure of particular diseases. When I name such men as Sir John Fife, Dr. Shepherd, Dr. Erasmus Wilson, Dr. Leared, and a host of others, I will be understood by physicians as referring to some of the leading therapeutists of the age; and, what is very interesting, they concur in their estimate of the bath, the thermæ, as a means of combating disease and fortifying health, with Hippocrates, the father of medicine—*Clarum et Venerabile Nomen.* These men all tell us that "in diseases of the skin, joints, liver and kidneys, the action of the Turkish Bath is immediate and direct," exercising an influence at once beneficial and powerful, and that in rheumatism, both acute and chronic, lumbago, neuralgia, sciatica and gout," the relief afforded is almost incredible. I can not forbear quoting one case, reported as occurring at the Newcastle Infirmary, by Sir John Fife:

"One of the worst cases of rheumatic gout that I have seen was admitted into the hospital October, 1860—a baker, aged 46. Since 1855 he had suffered from the affection in his joints and had been under treatment in two hospitals without relief. His elbows, wrists, fingers, knees, and ankles, were much enlarged, and stiffened so as to cripple him. In February, 1861, he thus expressed himself: 'Before coming here I had for the last two years been getting much worse. During the whole of this period I had very little refreshing sleep, and had continued gnawing, acute pain in all my affected parts. I have experienced much benefit from every bath I have taken, and can now use my hands and arms with much freedom. Previously I could not stand alone, and now I can walk without assistance.'"

This is very striking, but what follows is still more so. Mr. Urquhart tells us that one morning an "old man" (he does not state the age), who had for fourteen years been a victim to rheumatic gout, with thickening of the joints, and who, in all that time, had not been able to stand upright, went to the baths and remained there for an hour and a half—the heat at 145 deg.—and was so much relieved that he came out supported by two men. Mr. Urquhart met him as he came out, was struck with the appearance of the patient, who was a man of immense frame, and induced him to return with him into the bath. He raised the heat to 170 deg., and for the rest I must quote his own words: "I shampooed him, not in the ordinary way, but applying my whole strength, hitting him as hard as I could, and standing on his chest and limbs. In the intervals I subjected him to alternate rushes of hot and cold water, as he lay flat on the floor to get its full weight. After three hours he walked away erect. The chain of fourteen years was broken in a single operation, His own expression was: 'I went in on all-fours, and went away on wings.'"

There is another still more terrible because more fatal disease than rheumatism, whose ravages are now far more wasting than in former days. I mean pulmonary consumption. What I am about to say on this subject is drawn entirely from what I have read; but I am bound to say that there is much plausibility, at least much appearance of truth, in what is said by the advocates of the Turkish Bath in Great Britain, respecting the capacity of the treatment there administered to control this dreadful malady. Remember that the speakers are medical men, and that they deliver their testimony before a critical and vigilant profession. These men—some of them are among the most respectable and eminent of their profession—declare confidently that consumption may be cured, not merely alleviated, but cured, if, at the time the patient is brought within the operation of the thermæ, there remains enough of the substance of the lungs to carry on the process of respiration. It is not pretended that a waste beyond this point is capable of cure, but it is claimed that "the lesion of the lung may be arrested and healed," provided this condition obtains; in other words, that the current of decay can be checked, that what is not already swept away may be saved, and that an effectual stop may be put to the tendency to death. I said before that I only repeated what is reported by others. Relata refero. I admit the strangeness, almost the incredulity, of the statement. But it is supported by what seems irrefragable evidence. Several cases are reported in which persons given over by their medical advisers, and hopeless themselves, have been restored to health. And, paradoxical as it may seem, the strongest case is one in which death ensued; for the fatal termination was due to an accident, and a post mortem examination disclosed the fact, already confidently announced by the medical man in charge of the bath, that there had been extensive ulceration of the lungs and that this ulceration had under this treatment almost entirely disappeared. The patient was killed by an accident before he was discharged as completely cured, and thus, by his autopsy, may be said to have furnished more convincing testimony of the efficacy of the treatment than could have been possibly given if he had continued to live.

It is part of my disposition to enter with much earnestness upon whatever I wish to promote, and it would be easy for me to accumulate cases, to collect what lawyers call a mass of cumulative testimony, in support of the general proposition that the Turkish Bath is an agent of almost incalculable good, both in the prevention and cure of disease. But you would tire of my illustrations before I, in my zeal, would perceive that I had abused your patience. If what I have said awakens no attention and creates no disposition to examine into the merits of this great therapeutic power, I have already said too much. If I have been instrumental in promoting the establishment of the Turkish Bath in St. Louis, it will be impossible to deprive me of the consciousness of having rendered to her citizens a service the benefits of which will continue to be felt long after I am as much lost to memory as if I had never lived.

THE TURKISH BATH.

MESSRS. EDITORS: We desire, through the medium of your journal, to call the attention of the profession to the Turkish Bath as an important and powerful remedial and curative agent, and worthy of a more general application in the treatment of disease than has hitherto been accorded it.

Since opening a "bath" in this city we have given more than five thousand baths, and number among our patrons many of the best families in the city—clergymen, doctors, lawyers, merchants, tradesmen, etc.—men, women and children. A large majority of persons use the bath as a luxury or as a sanitary measure. As a luxury, all who take the bath concede that there is nothing can excel it; as a means of personal cleanliness, there is nothing can compare with it. But we have also treated with it quite a variety of diseases with, we may say, remarkable success. Fever and ague has often been cured with from one to three baths—in some cases, however, a bath daily was required for a week or ten days; and, so far as we know, there has not been a single re-occurrence of the disease in a single instance. Chronic diarrhœa of several years' duration has been completely cured in a few months by the constant use of the bath. Acute diarrhœa, cholera morbus and dysentery have been cured with from one to four baths. Neuralgias that had been treated in every imaginable way previous to coming to the bath, have yielded readily to its use. Rheumatism, both acute and chronic, is completely cured by it, as hundreds of cases would be pleased to certify. Spasmodic asthma is relieved by it a short time after entering the hot room. A number of young persons of both sexes, who, while growing very fast, seemed disposed to tuberculosis, after using the bath from week to week, have gained flesh and strength, and now enjoy robust health. Children from a few months to fourteen years of age have been brought in with a variety of ailments, and have been benefited in almost every instance. Not a day passes that persons do not express their high appreciation of the benefits they received from the bath. Indeed, we regard the bath as one of the most potent modifiers of the human organism, whether in health or disease, and we believe that all that is requisite to give rank as a legitimate weapon, so to speak, in the treatment of disease, is a more thorough investigation of, and general familiarity with, its effects by the profession; and to facilitate this, we offer physicians free access to the "bath" whenever it may suit their convenience, and desire that they bring with them one or more patients in order that they may watch the effects from day to day.

We would also suggest the introduction of a Turkish Bath in one of the hospitals of the city—the expense would not be great—and we would cheerfully give, without charge, the benefit of our time and experience in order that it might receive a thorough trial.

The Turkish Bath is now used as a remedial agent in nearly all hospitals in Ireland, and many in England and Germany, and the results, if we may place any reliance upon the home surgeons' reports, have been highly satisfactory.

The first lunatic asylum that introduced the Turkish Bath was that in Cork, the largest, except one, in Ireland, five hundred patients being the average. Dr. Power, the resident physician, says: "After four months' use of it, I found that seventeen persons had been perfectly cured by it, and sent home to their friends. The cases to which I now allude were a long time in the house, and classified with the incurables. * * * * * * * Now from fifty to eighty patients are daily admitted to the bath, many for its remedial effects, but the greater number from motives of cleanliness. Even these latter are wonderfully improved in appearance, and have acquired the ruddy glow of health, instead of the pale and sickly glow of invalids. Its salutary effect has been of a marked character. It fulfills all expectation. Since the introduction of the Turkish Bath into this asylum our cures have been forty per cent.; before its introduction only twenty. The deaths also have been lessened more than one-half." Now, if the introduction of a Turkish Bath into an insane asylum has actually doubled the cures and lessened the deaths more than one-half, is it not time that the medical men of this country should examine into and introduce into their practice this greatest of health-giving and most potent of remedial agencies?

Dr. Robertson, the medical superintendent of the Sussex Lunatic Asylum, England, says: "Insanity is a disease depending on and associated with various functional disorders, and especially with the perverted nutrition of the organs of the mind. The treatment of the pathological conditions consists not in the mechanical administration of specifics, but in the rational application of the principles of medicine to each individual case. To illustrate my meaning by a case: A patient is suffering from an attack of mania, with restlessness and incoherence of thought, and violence, with increased action of the heart, and congestion of the head, suppression of the catamenia, and of the secretions of the skin, which is rough and dry. The indications of the treatment here are, to restore the balance of the circulation, and thus to regulate the secretions and the supply of blood to the brain, and so to restore the healthy action of the uterus, the skin and the brain. Experience teaches us that such a result will only follow the slow and steady use of remedies influencing the action of the heart and of the nervous system. Of such remedies none are so powerful or so certain as the Turkish Bath; and I find that the continued use in such case of this remedy will, through its soothing action on the nervous system, and the relief it affords to internal congestion by determining the blood to the surface, modify, if not cure, the symptoms of the disease. In irregularities of uterine functions, which in young girls is sometimes complicated with mania, I have found in several instances a cure follow the restoration, through the agency of the bath, of the healthy uterine action. Setting the mental symptoms aside, I would here say that if the bath had only this one remedial power of restoring suppressed menstruation, its value in reducing the ills resulting from our high civilization would be still great."

Erasmus Wilson, an eminent authority certainly, says: "I thought I knew as much about baths as most men. * * * I knew their slender virtues and stout fallacies; they had my regard, but not my confidence." But after personal trial of the Turkish bath, he says: "I discovered that there was one bath that deserved to be set apart from the rest—that deserved, indeed, a careful study and investigation. The bath that cleanses the inward as well as the outward man; that is applicable to every age, that is adapted to make health healthier, and alleviate disease, whatever its stage or severity. It deserves to be regarded as a national institution, and merits the advocacy of all men, and particularly of medical men, of those whose special duty it is to teach how health may be preserved and disease averted."

Sir John Corregan, one of the most eminent physicians of Ireland, formerly a very bitter opponent of the bath, and who had written more against it than all the medical men in Europe combined, after a first trial of it for the relief of "phthisical laryngitis," from which he was suffering to such an extent that he compared the swallowing of any fluid, even, to "swallowing fire," or "running a red hot poker down the throat," said: "I have not experienced such relief for the last six months."

Spencer Wells, lecturer on surgery in the Grosvenor School, London, says: "One of the most common objections raised to the bath is the fear of taking cold in the transition from a heated room to the open air; but experience proves that this fear is groundless. * * * The habitual use of the bath tends to restore the normal properties of the skin, and leaves it less susceptible, and the bather may pass from the cooling room with impunity."

Dr. Watson, late physician to the Middlesex Hospital, England, in a lecture on diabetes, says: "There is another remedial measure which has a most beneficial influence on the condition of the patient. I mean forced perspiration, induced by the hot air bath. Of this I have seen some striking examples in the hospital."

Sir John Fife says of the bath: "Its effects are most remarkable in obviating disorders, and palliating diseases of the liver and kidneys."

Dr. Richard Barton says: "There is no power more capable of purging the blood than the bath, in cases where the pulse and stethoscope give unmistakable signs of disease of the heart; such patients take the bath with unlooked for benefits."

Had we space we could fill page after page with quotations from the best medical authority in Europe, but will merely add the following from Dr. John Balbirnie, a celebrated physician and author, who, in his "Plea for the Turkish bath," says: "If I were asked to give a brief and distinctive definition of the 'Turkish bath,' I would say, it is that which claims the exclusive or pre-eminent power of physiologically opening the body's safety valves; or, in other words, developing a high activity of the depurating organs; and so fulfilling the first grand condition of the cure of disease. If wielded by courageous and skilled hands, no artificial or medicinial system will be able to compete with it, either as respects the quantity or the quality of its cures."

In conclusion, we would renew our invitation to physicians to investigate the subject for themselves—to bring with them one or more patients, and watch the effects, feeling assured that a more general familiarity with the Turkish bath, by medical men, would tend to its more general recommendation and adoption in quite a large number of diseases.—[GEO. F. ADAMS, M. D., 1603 *Washington avenue, in the Medical Archives for September.*

SOME OF THE ADVANTAGES OF A TURKISH BATH OVER THE RUSSIAN BATH.

From the Family Herald, St. Louis.

It is often asked what difference is there between a Turkish and Russian or steam bath? and but few are fully acquainted with the relative merits of hot and cold water baths. The Turkish Bath stands by itself on its own peculiar merits; it is the result of the best scientific research in Europe. This bath is the *hot air bath.* Neither hot steam nor hot water is applied. The person receiving the Turkish Bath reclines at ease upon a couch, and in a few minutes the perspiration, no matter how long it has been checked, pours forth, and he experiences a sense of relief, comfort and ease that prove how valuable it is to the fatigued or sick. To the well person it is not less a luxury. It is curious to note the rise of the Turkish Bath in this country, in spite of all opposition; it is fast superseding the Russian. Every new bath now being established is Turkish. New York has eight Turkish Baths, six of which have been put in operation within the last three years, while only two Russian Baths are now in use in New York city. Brooklyn, N. Y., has three Turkish Baths and not one Russian. Philadelphia has two Turkish Baths and another in contemplation. Boston has three, and no Russian. Chicago has two, and Indianapolis one, in a private hospital. Milwaukee has one fine one, and San Francisco is to have a new Turkish Bath at once, at a cost of a quarter of a million of dollars.

All physicians who have given a thorough investigation of the subject, prescribe this bath in preference to any other. The success of this treatment has been so great and decisive that thousands resort directly to the bath when indisposed. There are many diseases of which it may be safely said, the Turkish Bath is sure to help or wholly cure. It expels morbid matter from the system and prevents disease; two or three baths often save two or three weeks' illness. Some of the peculiar properties and virtues of this bath, which render it so superior to the Russian or any other mode of treatment, are particularly stated by Dr. Spencer Wells, in his lecture delivered at Grosvenor Place School of Medicine, in London, England. He says:

"Until lately, vapor baths were the only means at our command for inducing a profuse perspiration, but the introduction of the hot air or Turkish Bath is a very important addition to our means of preventing and treating disease. Our bath, though unworthy of the idea that we *now* attach to the word bath, hitherto has been water or vapor; now we have hot air. You will see at once the great importance of this distinction when we reflect that in one case the body is surrounded by dry air, which must favor the exosmosis of the watery portion of the blood through the coats of the cutaneous capillaries, and the endosmosis of oxygen, and at the same time must favor evaporation; while in the other case, the body is either immersed in water or surrounded by vapor, which would be absorbed in place of oxygen, while evaporation would be checked. In the one case you have exosmosis of fluid and absorption of oxygen; in the other case you have neither."

Every Turkish Bath bather knows that dry air can be supported to a far higher degree of heat than air which contains much moisture, so that we can order baths of far higher temperatures than we ever thought of before. Hence, the confounding of the Turkish Bath with the ordinary hot water or vapor bath is perfectly absurd, or of comparing the results of the one with the other. *All this should be well understood by the medical men*, or they will find their patients know more about it than they do, and nothing can be conceived more damaging to their prospects of professional success. If the young physician hopes to succeed, he must keep ahead of his patients in knowledge of everything appertaining to their health. Now that the hot air baths are springing up in every city, the interest in them steadily increasing, the public are clamoring for information in regard to them. "How many, I ask, can give an intelligent answer to such inquiries?" It seems to the writer of this article that it is every physician's bounden duty to make himself thoroughly acquainted with the principles of the hot air bath, instead of turning off his patient with a—"well, you can take them, but I doubt whether they will be of any benefit to you; or that they are good in some cases, but are not indicated in your present condition; or that you are too weak, too full-blooded; or some weak and silly answer that does not satisfy the inquirer at all, and really places the physician in a false position, for there are thousands of our most enlightened and intelligent gentlemen and ladies *who know* what a Turkish Bath is, and what its results are when properly administered. It is of no use for those who know little or nothing in regard to the bath to say that it is a humbug, and attempt to keep people away from the bath by saying it very debilitating, and but few are able to endure the extreme changes from hot to cold, for there are no such changes. Look at the men and women employed in the bath from four to ten hours daily. Are they sickly? Are they weak and feeble? Let those who are regular bathers answer. The writer has seen men and women who have worked the baths for ten years and upward, and finer specimens of muscular development would be hard to find, even among the professional gymnasts. I may also add, that during the ten years so employed they have not lost a day's time by sickness. Such testimony as this can not be winced out of sight, and those who attempt to bring disfavor upon an institution so conducive to the health, happiness and well being of society, by misrepresentations, does so either from prejudice, selfishness, or some other damaging motives.

The bath used simply as a cleansing process, excels all others, for in no other way can that sense of cleanliness be realized. An eminent author has truly said, "the person who has never taken a Turkish Bath has never risen to the moral dignity of being personally clean." Finally, the writer himself believes that if every man, woman and child in this large, dirty city could avail themselves of one bath a week, much sickness would be prevented, many valuable lives prolonged, and many saved; in a word, it would enhance all the enjoyments of life.

The Turkish Bath challenges the attention of all scientific men as one of the most harmless and most powerful means of removing disease, while at the same time it is one of the most delightful luxuries imaginable.

PRO BONO PUBLICO.

WHAT TURKISH BATHS DO FOR A PERSON.

Those who have indulged in the luxury of Turkish Baths require no argument in their behalf. But for the benefit of the uninitiated, it is well to state that, of all the delicious sensations that ever thrilled through the mortal body with gentle fervor, none can compare with those excited by the dreamy influences generated therein. Words can not describe the sense of tranquil rest that soothingly captivates the weary body, and lulling the tired brain, brings about a mental equipoise that is the acme of delight. And yet the Turkish Bath has a higher use than that of mere sensational enjoyment.

It is a grand promoter of health, as it equalizes the circulation throughout the human system, eliminating from the body all impurities, and making it impervious to malaria and other ills. Having experienced its benefits, the writer would call attention to the establishment of Dr. George F. Adams, No. 1603 Washington avenue, where the bath is daily administered with Oriental exactness by attendants of European experience and fidelity. Although Dr. Adams has been located in this city but little over two months, baths have been given to over thirteen hundred persons, male and female—the former under his own personal supervision, and the latter under that of his wife. At this season of the year, when the changeable weather is productive of colds, the bath is a blessing. It removes a cold and prevents others, to say nothing of fever and ague, neuralgia, and other distressing complaints, and leaves the system less susceptible to sickness than before. Writing after personal experience, we can, with all honesty of purpose, advise those of our readers who desire to taste the joys of health and luxury combined, to visit the attractive home of Dr. Adams, and there revel in the supremest refreshment of which the body is capable.—*Missouri Democrat.*

The Turkish Bath---Its Character and Philosophy.

PARTICULARS OF THE WASHINGTON AVENUE ESTABLISHMENT.

A DESCRIPTION OF THE MODUS OPERANDI OF THE BATH.

Its Delightful Sensations—A Glance at the Process—Its Effects on the Physical Frame—The Popular Delusions Respecting the Bath—Opinions of Eminent Physicians on the Subject—Testimony of St. Louis Citizens, &c., &c.

It is now a little over a year since the Turkish Bath Establishment of Dr. George F. Adams, at No. 1603 Washington Avenue, commenced operations. Its career during the period has been most successful and satisfactory, and it may now be regarded as one of the permanent institutions of the city. In this article we do not propose to go into any minute description of the Bath, but to draw attention to its valuable character as a curative agency and a health-preserving luxury, as practical experience has demonstrated the fact.

The Washington Avenue Establishment.

Before presenting some authorities and particulars on this subject, it is but just to state that the establishment of Dr. Adams is so complete in its arrangements and conducted so admirably that it fairly presents the famous Turkish Bath for candid examination. As exemplified in this institution, the principles of hygiene involved in the Bath can be scientifically tested and their results fully demonstrated and developed. Dr. Adams is a physician of standing in the profession, and has spared neither trouble nor expense to perfect his establishment. It is the Turkish Bath pure and simple, and not as marred and perverted by ignorance and blundering management. When we say the Turkish Bath, however, we do not mean as it is generally carried out in the Orient, for there it is very much inferior to the improved process which characterizes the best baths of Europe and America. The East must be credited with the fundamental principle of the bath, but this having been once received by the higher cultivated genius of the West, it has been expanded materially and its practical application highly improved. It has been aptly said "Western Europe was indebted to the East for the Bath, but the East has to thank the West for its proper construction and manifestation of its curative power."

The Turkish Bath—What is it?

For the benefit of some of our readers who may have no practical knowledge of the subject, or entertain some erroneous impressions respecting it, we will answer the question. The Bath consists of four rooms, and, gentle reader, let us undress and go through them in proper order. In a small, neatly furnished compartment, shut off by a curtain from the cooling room, we prepare for the bath in perfect seclusion. We finally come forth in a state of nature excepting a crimson cincture. We are ready for the bath. An attendant pulls a bell, opens a door and we descend some steps; another door opens and we enter the calidarium. In a moment we are in a region of intensified summer air. The usual shrinking sensitiveness of the unclothed frame vanishes. The warm soft air seems to fold us in a delicious embrace. All apprehensions about the Turkish Bath disappear in this safe, delightful place. The attendant motions us to one of the comfortable reclining seats, over which is spread the drapery of a snowy sheet. We lie down, and the air, the exquisite, caressing, warm air, invades us everywhere and the luxury of the passive enjoyment increases every minute. We glance at the thermometer; it is 140 degrees, and we wish it was even higher. Pain, weariness, languor and sick feeling have disappeared, and, like the mystic lotus eaters, we rest in tranquil, drowsy delight. Ten minutes pass as we lie gazing at the sky through the ceiling window, and the real world is half hidden. The imagination grows active, and it requires but a slight effort to supplement the exquisite touch of the temperature, and we are in the land of eternal summer, with the bliss and beauty of perfected nature around us. Now the skin grows moist and glistens with its exudations. The face, the head, the hands, neck and shoulders perspire first, because in the most normal state from exposure to the air. Soon we are "like Niobe, all tears," but not of pain, and even as the perspiration grows more profuse there is no feeling of weakness, because it springs from no physical effort. We feel buoyant, and happy, and disposed to mirth. The attendant examined us critically, and then leads us into the warmest division of the calidarium. This is on the side at which hot air is admitted, and it is separated from the one we have left only by a curtain. The difference in temperature is, however, ten degrees. We take another couch, and now the heat is 150 degrees. We enjoy it the more, and renew our dreams.

It must be understood that this hot air is perfectly pure, and so tempered by passing over cold air tubes, as to be rendered perfectly soft. It is absolutely sweet to breathe, and those who only know hot air as it flies oft rusty steam-coils, have no idea of what it is here. After half or three-quarters of an hour or more has been passed in this chamber, we are led into the shampooing room. Here a polished marble slab, warm as the air, receives our recumbent frame. Now .commences that gentle, but wonderful washing, which, until a person has taken a Turkish Bath, he can have no idea of. A fragrant lather envelopes us, while soft brushes glide over us. The inward man was cleaned in the calidarium, and now the outward man is made as pure as Adam when he first opened his eyes on Paradise.

The free application of water succeeds this by means of the spray, douche and shower bath. At first it is warm, but the temperature is graduated, shading into tepid and cold almost imperceptibly. There is no abrupt transition, no violent change, and extreme cold water is only turned on at the bather's request. Most persons, however, desire it, for the gradual process of the bath has so educated him, that he stands the dash of an ice-cold stream without a shudder and with positive pleasure. The exuvia thrown out from the system is removed, the pores are closed and the skin tempered. The water bath commences just high enough to avoid chilling, and the temperature lowers gradually to that of spring water. It must be remembered that during these steps in the bath the bather is completely passive, and hence emerges at the end not only not weakened, but full of exhilaration and vigor. He is now dried and enveloped in a sheet, re-enters the cooling room and takes a siesta on a couch, smokes a cigar, reads or sleeps for half an hour or an hour. Then he is ready to dress and depart into the open air.

Effects of the Bath.

Many delusions have existed, and still exist, respecting the Turkish Bath, and a person never rightly understands it until he has practically tested its virtues. A common idea is, that it has a weakening effect and renders one liable to cold. This is entirely fallacious. Instead of consuming strength, it develops it; and on emerging into the open air after having gone properly through its process, one feels fortified against cold. The profuse perspiration produced without exertion has cleansed the system, and a normal temperature is regained by a graduation that does violence to no organ or part of the frame. Vigor of body and mind, and an elastic cheerfulness are generally the immediate effect of the bath, and more particularly where it is repeated two or three times a week or oftener.

Medical Opinion on the Matter.

In referring to medical authorities for an expression of opinion on the subject, we find a vast mass of authority unanimously in its favor. There are, of course, some physicians who hesitate in indorsing the Bath: they are entirely among those not personally and practically acquainted with its character. We will give some idea of the medical approval which supports it. John Balbirnie, an English physician of eminence, a member of the Royal College of Surgeons, and author of a number of medical works, has also published a pamphlet of some forty-five pages on "The Physiological Basis and Curative Effects of the Turkish Bath." In the preface to the first edition, after characterizing it as "a pure custom, a new mode of cleanliness," &c., he adds in italicised letters, "*This bath is also a mighty agency for the prevention and cure of disease.*" At the close of his treatise he enumerates twenty-nine cases in which the bath acts either as a direct cure or a preventive agency. He further says: "The first essential element of the action of the Turkish Bath is hot air, the purer the atmospheric oxygen and the freer of all admixture of dilution, clearly the better. Under this stimulus the whole secretory activity of the system is aroused, transpiration is powerfully increased, both from the skin and lungs, with the effect of imparting extra activity to the circulation. Every vital, vegetative or purely organic function is stirred up to unwonted activity, the heart beats with renewed energy, and the blood vessels participate in its augmented impulse, with increased outpouring of the structural debris—veritable body-sewage—unhealthy elements imprisoned within are loosened, set afloat, and swept off by this real flood-tide of fluids speeding onward to the surface, like rivers to be lost and exhaled in the ocean. The completeness of the æration of the blood corresponds in degree to the activity of exhalation, respiration is deepened and the lungs are profoundly filled."

Dr. Fisher, M. R. C. S., says: "After having had many Turkish baths, I feel more and more convinced after each bath that we have nothing equal to them as eliminators of noxious matter from the human system. In these days we do not take sufficient exercise to make us perspire freely, hence this artificial mode of inducing perspiration is useful, particularly for those whose habits and occupations are of a sedentary nature."

Dr. Erasmus Wilson, F. R. S., says: "The Turkish Bath may be ranked among the very foremost of the necessaries of life. To remove all impurities from 7,000,000 pores is to cleanse and ventilate twenty-eight miles of drainage, and this is accomplished every time these baths are taken."

Speaking in connection with the bath, Dr. Mitchell, in his "Therapeutics," states: "Every practitioner knows that he cannot excite perspiration on the first day of a high, burning fever by any sudorific as well as after the action has been reduced by blood-letting and vomiting. Quite a severe preliminary this to go through to secure the action of the medicine. Heat, I affirm, is the only certain, direct diaphoretic applicable to the preservation of health and cure of disease, and should be employed by every one who undertakes the guardianship of the health of his fellow-beings, rather than tamper with other uncertain and inefficient agents."

Dr. R. M. Lackey, late Demonstrator of Anatomy, Rush Medical College, has published a pamphlet on the bath, advocating its general use as an invaluable agent in promoting health and curing disease—particularly for rheumatism and neuralgia, diseases of the kidneys, bronchitis, asthma, consumption, catarrah, and that class of human ills; also dropsy, diarrhœa and dysentery, fevers, bilousness, colds. In reference to the use of the bath in insane asylums, it is stated in Great Britain, where it has been thoroughly tested, it is shown that the addition and use of the Turkish Bath in insane asylums will double the number of cures, and reduce by a half the number of deaths.

The Report of the Lunatic Asylums of Ireland says: "In the Cork Asylum, where the Turkish Bath was used, the number of patients discharged cured, during the two years, was forty per cent. of the average number, while in other asylums in Ireland, where the bath was *not* used, the number discharged was but twenty per cent."

Dr. Robertson, of the Sussex Lunatic Asylum, remarks: "The bath realizes more for me than any other agent the requirements of rational medicine."

Mr. David Urquhart, who was mainly instrumental in the introduction of the bath into England, and who gave to the subject the most laborious examination, writes: "There is an impression among those who never have tried it, that the bath may be weakening. We can test this in three ways: First, by its effects on those debilitated by disease; second, on those exhausted by fatigue; and third, the attendants of the bath. First—In Turkey, in affections of the lungs and intermittent fever, the bath is invariably had recourse to against the debilitating night sweats; the effect is to subdue by a healthy perspiration, in a waking state, the unhealthy one in sleep. No one ever heard of an injury from the bath; the moment any one is ailing he is hurried off to it. Second—After long and severe fatigue, successive days and nights on horseback, the bath affords the most astonishing relief. A Tartar having but one hour's rest prefers a bath to sleep. Well can I remember the haman which I have entered, scarcely able to drag one limb after the other, from which I have sprung into my saddle elastic as a sinew and light as a feather. Third—The shampooers spend eight hours daily in the bath, and are remarkably healthy. They enter the bath at eight years, and the best shampooer under whose hands I have ever been was a man whose age was given me at ninety."

Dr. Baxter states: "You see young girls snatched off by consumption; you see people with scars and marks on them from scrofula. Now, it is a curious fact, that where the bath exists in the East scrofula is unknown, and gout and rheumatism are medical curiosities."

Sir John Fife, M. D., gives his testimony thus: "Its effects are most remarkable in obviating disorders, and palliating diseases of the liver and kidneys."

We might indefinitely increase the number of these medical expressions of opinion, but enough is furnished to indicate that the subject is one which has been thoroughly examined, and that the bath is commended by high authority. Among the English physicians who have spoken highly of it are Sir H. Marsh, Physician in Ordinary to the Queen; Thomas Watson, M. D., late physician to Middlesex Hospital; R. H. Golden, M. D., physician to St. Thomas Hospital; John Forbes, M. D., late physician to the Queen's household.

PATRONAGE BY ST. LOUIS CITIZENS.

Apart from the sanction of medical authority, the ordinary healthy visitor who visits the bath carries away a witness in himself. He feels better every way, and has a consciousness of strength and endurance that is new and delightful. In St. Louis Dr. Adams has a multitude of testimonials from prominent citizens who regularly visit the bath. Among these names are the following: T. G. Comstock, M. D., Frank G. Porter, M. D., J. S. Read, M. D., Wm. M. Barker, M. D., Wm. Helmuth, M. D., S. J. Brackat, M. D., G. L. Walker, M. D., Montrose A. Pallen, M. D., I. G. W. Steedman, M. D., Thomas O'Reilly, M. D., I. M. Franciscus, D. A. January, John R. Shepley, Chas. Whittlesey, R. B. Whittemore, P. W. Hermans, George I. Barnett, Frank P. Blair, James L. Knight, R. S. Voorhis, Henry Hitchcock, Jos. O'Neil, S. D. Easter, E. W. Fox., H. S. Turner, T. T. Gantt, Geo. D. Hall, W. L. Pottle, L. M. Dodd, Albert Todd, H. W. Jones.

We may add, in conclusion, that the establishment of Dr. Adams is elegantly furnished, and all the appliances of the bath are kept constantly in excellent order.

TESTIMONIALS FROM PATRONS OF THE BATH.

SOUTHERN HOTEL,
St. Louis, March 29, 1871.

DR. G. F. ADAMS—*Dear Sir:* Having tried the Baths of Wiesbaden, the Hot Springs of Arkansas, and various other kinds of "electric baths," I feel assured no bath will afford greater relief to the rheumatic than the Turkish Bath, and would advise ALL to give them a trial—the afflicted for the benefit to be derived; the healthy for the luxury afforded. Yours truly, JOHN H. LOUDERMAN.

ST. LOUIS, April 15, 1871.

DR. GEO. F. ADAMS—*Dear Sir:* The "Turkish Bath," as given at your establishment, should be taken to be appreciated. I have taken the baths in other cities, and it gives me pleasure to state that, for completeness in detail and competent and polite attendance, yours is superior to any that has come under my observation.

Yours truly, JOHN G. PRIEST.

PEEKSKILL, New York.

DR. GEO. F. ADAMS—*Dear Sir:* I am sorry that you are going to leave Brooklyn; but if it is best for you I can not complain. You have been sole physician in my family for seven years or more; your treatment has been highly satisfactory during all that time. We were more than satisfied, and feel that we are in your debt. You will remember in some cases, particularly in my wife's case, the circumstances required great courage, skill and common sense, and we felt at the time, and since, that you proved equal to the emergency. I can especially recommend you in the department of midwifery, where courage, steadiness and delicacy are so much required.

I suppose your three years' experience in the army have added largely to your surgical skill, but of that I have had no knowledge, except the favorable reports that have come to me. I wish you much success, a long and useful life, and so a happy one.

I am very truly your former friend and patient,
HENRY WARD BEECHER.

ST. LOUIS, November 6, 1870.

The Turkish Bath is a luxury that must be tried in order to be appreciated. It rejuvenates the physical system, and gives new life to man from the crown of his head to the soles of his feet. I regard the Bath as an indispensable luxury, and one of the greatest blessings that a man can confer upon himself. I would sooner deprive myself of any luxury that I enjoy than be without my regular weekly Turkish Bath. J. S. HAY, St. Louis Dispatch.

Dr. Adams—*Dear Sir:* I take pleasure in adding my testimonial as to the efficacy of the Turkish Bath. Having suffered some time with rheumatism, the Bath was recommended. "I came, I saw, I conquered." From the very first I improved. After the third bath that excruciating pain was gone, and I could walk with ease and comfort. When I commenced I could scarcely move about; could not sleep. A. V. Bohn,

Supt. Coal Co., 311 Pine street, St. Louis.

Boston, May 29, 1869.

Dr. Adams—*Dear Sir:* It is taking the bread out of my mouth, but I must tell the truth. The Turkish Bath is a great remedy, and those who take them are not among my best paying patients. If I recommend sick persons to take the Baths I have but little to do with them, professionally, afterward. G. M. Pease, M. D.

St. Louis, Mo., April 15, 1871.

I have been afflicted with palpitation of the heart for ten years, and have tried all the remedies known to me in the practice of medicine for upward of thirty years, and for the last three years I have tried bathing—baths at the sea shore and all the watering places in Virginia and in Pennsylvania—without finding any relief, save the Warm Springs in Virginia, which was only temporary. Last fall I commenced taking the Turkish Bath in this city, and only took them occasionally through the winter, and found so much benefit from them that I have been taking them regularly this spring—twice each week; and I can truly say that I have suffered much less from palpitation of the heart during the last winter and spring than I have at any time during the last ten years.

Jeff. F. Jermaine, M. D.

The Turkish Bath—A Healing Balm to the Suffering and a Shield to Health.—Among our various medical remedies for producing diaphoresis, hot air is the most potent; and how different and agreeable its mode of action when compared with drug diaphoretics, such as antimoni, guaiacum, ipecacuanha, etc. The skin is the only organ given to our care. We have eyes to see it if it is not right; we have hands to feel and noses to smell, and still it is the organ most neglected of all others. All the internal organs are guided by a beautiful internal involuntary organization beyond our control; and it shows the wisdom of such design, for if it were left to our care, like the skin, they would meet the same fate, and death would soon overtake us. I should be no friend to humanity nor to medical science if I did not give my testimony in its favor. One word more as to the Bath's remedial effects: My little daughter, aged three years and six months, was attacked with "chorea"—St. Vitus dance. The usual remedies were used for several weeks without any apparent benefit; then electricity, with no better success. I then submitted her to the Bath, with the most gratifying results. After she had taken four baths she was well. Three months have now elapsed and there has been no return of the disease.

SAMUEL J. BRACKETT, M. D.

St. Louis, March 10th, 1871.

ST. LOUIS, May 29, 1871.

DR. GEORGE F. ADAMS—*Dear Sir:* I cheerfully comply with your request, to furnish you with a statement regarding my individual as well as professional knowledge of the results attending the use of the Turkish Bath.

In the spring of 1870 I was attacked with Nephritis Albuminosa (commonly known as Bright's disease of the kidneys), which lasted several weeks. Following my *comparative* recovery from which, I suffered till the month of August with catarrh of the mucous membrane of the air passages, so that I was reduced to but litttle more than a walking shadow—in fact, weighing but 113 pounds. I was not a little surprised one morning of the entire absence of my usual copious expectoration, from which time, up to the 10th of October following, I rapidly and enormously acquired flesh (?), pulling down the scales at 208 pounds. During the time I was acquiring this enormous (for me) bulk, I never felt better in my life, until a casual remark, by a friend, suggested the propriety of my making the usual tests for DROPSY. I take from notes made at that time the following memoranda:

URINE—Sp. gr. 1007 (should have been, if healthy, 1026) surcharged with fatty epithelium scales; but *slightly* albuminous, and largely diffused with fibrinous shreds.

I assure you, Doctor, a result like the above was not at all pleasant to contemplate. Since the text-books are so incomplete as to the proper procedure in such cases, and, upon reflection, I marked out a faint line for my guidance, viz.: *To keep the emunctories open by the Turkish Baths*, at the same time supporting the system by the use of Ferruginon's Preparations, such as Ferri cit., F. sub-carb., F. pyrophos. and F. et quin. cit.

I took the bath three times a week for one month, and one to two a week the second month. From the very first I experienced absolute benefit, my perseverance in the same resulting in positive and complete recovery.

From the circumstances of my case I am free to declare: First, that the Turkish Bath, *if persevered in*, will cure dropsy, for I attribute the curative effects to it, having only supported the system with tonics while taking them for a primary object. Second, the baths are the greatest luxury of the nineteenth century.

Several patients of mine who have taken the baths at my direction have experienced like beneficial results.

In conclusion, I trust that the medical profession will early appreciate the Turkish Baths as a remedial agent in disease as it is a luxury in health. I am, Doctor, very truly yours,

R. MORRIS SWANDER, M. D.
1015 Cass Avenue.

The undersigned having enjoyed the pleasures and advantages of Dr. George F. Adams' "Turkish Baths," take great pleasure in recommending them to the St. Louis public, as complete in all their appointments, and well worthy of the patronage of those who wish to enjoy a real luxury, or to get relief from the many ills that these baths are capable of removing. The ladies who visit the bath will receive from Mrs. Dr. Adams every attention requisite to insure comfort and the best advice; while the gentlemen visitors will secure all the attention that they may require. In fact, no one can visit these Turkish Baths, 1603 Washington Avenue, without being delighted with them for their many advantages.

T. G. Comstock, M. D.,
Frank G. Porter, M. D.,
J. S. Read, M. D.,
Wm. M. Barker, M. D.,
Wm. Helmuth, M. D.,
S. J. Bracket, M. D.,
G. L. Walker, M. D.,
Montrose A. Pallen, M. D.,
I. G. W. Steedman, M. D.,
Thos. O'Reilly, M. D.,
I. M. Franciscus.
D. A. January,
Jno. R. Shepley,
Chas. Whittlesey,
R. B. Whittemore,
P. W. Heermans,
Geo. I. Barnett,
Frank P. Blair,
Jas. L. Knight,
Henry Hitchcock,
D. Robert Barclay.
J. O'Neil,
J. D. Easter,
E. W. Fox,
H. S. Turner,
T. T. Gantt,
G. D. Hall,
M. L. Pottle,
S. M. Dodd,
Albert Todd,
H. W. Jones,

[Contributed.]

Turkish Baths versus Arkansas Hot Springs

Much has been written, and more said, in regard to the invaluable properties of the Arkansas Hot Springs as a remedial agent. Invalids, from all sections of the country, flock to the Hot Springs with every variety of disease, and I am informed by the resident physician that all are more or less benefited, and many radically cured, where all other known means had failed. The question very naturally arises in the minds of those who are willing to search for new truths—What brings about these wonderful cures? It is evident to all who think for themselves, that it is not the mineral properties of the spring water, for the analysis of the water shows it to be as pure as distilled water, or nearly so, notwithstanding they have the arsenic, sulphur and alum springs; consequently, we must at once come to the conclusion that it is the "diaphoresis," or sweating, caused by the heat of the water, rather than by its properties. Of course, change of air, change of diet, freedom from care and business perplexities have much to do with the cures, yet facts go far to prove that the sweating process is the great secret of the cures. The largest class of drugs used by the profession are known as diaphoretics or sudorifics, and are used more generally than any other medicines, but at the same time they are the most uncertain in their operation, and, I may safely add, the most pernicious. Now, I claim (and my experience will bear me out in what I say) that heat stands mountains high above any and all diaphoretics as a cure in all kinds of diseased action, for precisely the same results can be brought about with heat without the injurious, and sometimes fatal, effects of diaphoretics. Now, if moist heat or steam is capable of equalizing the circulation, removing obstructions, relieving congestions, stimulating the circulation, and calling into action the secreting and excreting organs, how much better is the application of the pure dry heat of the "Turkish Bath." When I speak of the "Turkish Bath" I do not include those so-called Turkish baths gotten up by every adventurer, dubbing themselves Professors, Doctors, or what not—not knowing the first principles of physiology, or the laws of health. A bath with intense heat, poor ventilation, is no Turkish bath, and the public should not judge *the* Bath by such shams. Dry heat can be endured at a much higher temperature than moist heat, while steam or hot water can only be endured at 110° or 115° at the highest; the hot air bath can be taken as high as 150° with comfort, and many take it much higher. I have taken one in New York at 250° with the happiest results. Hot air favors evaporation; moist heat does not. Dry heat favors absorption of oxygen; moist heat prevents it. Dry hot air invites the blood to the surface, causing profuse perspiration; steam heat condenses upon the body and prevents free perspiration. A person with ordinary judgment knows this, by the acceleration of the heart's action. In a steam bath, many really think they are sweating, when it is nothing but the condensed steam pouring off of them. The Hot Air or Turkish Bath of our city commends itself to all, especially to invalids. I hold that it is, physiologically speaking, far ahead of the Arkansas Hot Springs as a remedial agent, in any and all diseases, for the reasons above stated. Give the Turkish Bath half the chance you do the Arkansas Hot Springs—that is, by attending to them as you would do when you go to the Springs, and you will find that my words will more than prove true. You will then save a hard and expensive journey, will retain all your home comforts. The accommodations of the bath in our own city are far superior to those in the wilds of Arkansas, and last, but not least, a competent physician, of long experience in the practice of medicine, gives his entire time and advice to his patrons, and his accomplished wife her time, during the ladies' hours, free of charge. We hope, with Dr. Adams, soon to see a Turkish Bath introduced into our city hospital and insane asylum. Who will be the happy man (physician or layman) that will immortalize his name by being instrumental in bringing about a consummation so devoutly to be wished? A FRIEND OF THE BATH.

The Turkish Bath in Diseases of Women.

I could not let this work go to press without saying a word on this important and delicate subject. Feeling assured that I have in the "Bath" an instrument of great power for good in all female complaints, I shall venture a few plain, truthful words to the ladies of St. Louis and vicinity, knowing, however, full well that I shall be misjudged by some and abused by many for doing what I know to be my duty—*my duty*, because I *know* that the bath will accomplish much to relieve and restore to health those who are now suffering from suppressed, painful or excessive menstration, chlorosis and leucorrhea. Many cases of prolapsus ani, as well as prolapsus uteri, I know have been radically cured by the remedial effects of the bath, without mechanical means. Nervous diseases, hysteria, St. Vitus' dance, nervous headaches, are often promptly cured by a few baths. Constipation, the bane of human happiness, so common among young girls, and so prolific of mischief—the chief cause of nearly all of the above diseases, is usually cured by a course of baths and proper attention to diet. For the complexion no cosmetic ever compounded will compare with the hot air bath; the skin becomes fair and soft, moth patches disappear, and that sallow, sickly look of the skin gives place to the ruddy glow of health. All this, and much more, is accomplished by a frequent use of the bath. I know too well what many will say to these bold statements, but truth is stranger than fiction. It is more than probable your physician will, for the present, discourage you from trying them, but the time is not far distant when they will be recommended by all true physicians. A want of knowledge as to the true merits of the bath as a remedial agent is all that prevents most of our physicians from accepting them as orthodox, or rather as a legitimate remedy. Having made obstetrics and the diseases of women almost a specialty for more than eleven years, in Brooklyn, N. Y., previous to entering the army, I feel that I have a right to express my opinions freely on this subject, and I claim some little indulgence in the means I may use in bringing the subject fairly before the ladies of St. Louis.

There are many other topics that might be discussed under this head, but I will close by a word of advice to those who may have the good sense to try the baths, even against the advice of their friends and physician: 1st. Consult those who have tried the bath—those who know something about it. 2d. Do not judge of the effects of the bath until you have given it a fair trial, for it is frequently the case that one or two baths will disturb the equilibrium of the system, and make you feel as if you were under the operation of medicine. 3d. Remember there is no other known means that compares with the bath in equalizing the circulation; consequently obstructions are removed, the secreting organs are kept active, nutrition stimulated, congestions removed, and many diseases entirely controlled by this great instrument—the bath. I say this on the authority of four years' experience, having given it my undivided attention during that time, also on the authority of such men as Dr. H. Marsh, Physician in ordinary to the Queen of England; Dr. Thos. Watson, late Physician to Middlesex Hospital, England; Dr. R. H. Goolden, Physician to St. Thomas' Hospital, England; Dr. John Forbes, late Physician to the Queen's household, England.

PUBLISHER.

ERRATUM.—The article on page 57, headed "Chapter VII," is a separate article, and should have been credited to DUNHAM DUNLOP, M. R. I. A., being an extract from his work entitled, "The Bath; or, Air and Water in Health and Disease."

TURKISH BATH DIRECTORY.

NEW YORK CITY.—Dr. Angels, corner of Twenty-fifth street and Lexington Avenue; Dr. Miller, 41 West Twenty-sixth street; Dr. Woods, 15 Laight street. There are several others; but the above I am personally acquainted with, and can cordially recommend them.

BROOKLYN, N. Y.—Dr. Shepard, 81 & 83 Columbia street, Brooklyn Heights. A most excellent Bath. A Mr. Burnham, also, has a fine Bath on Smith street.

BOSTON, MASS—Has several Turkish Baths, one at 1427 Washington street; one under the Malboro Hotel, Washington street; Mr. Fields, No. 18 Harvard street, just out of Washington street. This is a charming little Bath, well kept and centrally located. Travelers will be well pleased with all its appointments; and he understands his business thoroughly.

PHILADELPHIA, PENN.—Has two excellent Baths. I am not posted as to their location.

CHICAGO, ILL.—Has two—R. M. Lackey, M. D., 294 Wabash Avenue, and one on or near the corner of State and Washington streets.

MILWAUKEE.—M. P. Hanson, M. D., corner of Fourth and Sycamore streets. Parties traveling East will find the above Directory very useful.

PUBLISHER.

www.ingramcontent.com/pod-product-compliance
Lightning Source LLC
Chambersburg PA
CBHW032247080426
42735CB00008B/1044
* 9 7 8 3 3 3 7 2 9 0 7 9 5 *